VIDEOFREEX

America's First Pirate TV Station & the Catskills Collective That Turned It On

by Parry D. Teasdale

BLACK·DOME
RR 1, Box 422
Hensonville, New York 12439
Tel: (518) 734-6357
Fax: (518) 734-5802

Published by

Black Dome Press Corp.
RR1, Box 422
Hensonville, New York 12439
Tel: (518) 734-6357
Fax: (518) 734-5802

Copyright © 1999 by Parry D. Teasdale

All rights reserved. No part of this book may be reproduced
in any form except by a newspaper or other reviewer who wishes to
quote brief passages in connection with a review.

Library of Congress Cataloging-in-Publication Data:

Teasdale, Parry D.
    Videofreex / by Parry D. Teasdale
    p. cm.
    ISBN 1-883789-21-4
    1. Videofreex (Production company)—History. I. Title.
PN1992.945.T43 1999
384.55'06'573—dc21   99—27125
                                                      CIP

Cover photo: This formal portrait of the Videofreex in 1973 appeared on the back cover of *The Spaghetti City Video Manual*. Front, from left, Ann Woodward, David Cort, Murphy Gigliotti (Davidson's son), Carol Vontobel, Nancy Cain, Davidson Gigliotti. Rear, Sarah Teasdale, the author, Skip Blumberg, Bart Friedman and Chuck Kennedy.

Chapter illustrations by Ann Woodward
Design by Carol Clement, Artemisia, Inc.
Printed in the USA

Ann Woodward's illustration for the cover of *The Spaghetti City Video Manual* depicts each of the Freex and their pets in 1973, when the book was published. From left, they are: Davidson Gigliotti, David Cort (with Oberon on his shoulder), Skip Blumberg, Ann Woodward (foreground), Carol Vontobel, Chuck Kennedy (Mushroom is behind him), Sarah Teasdale (foreground), Nancy Cain, Bart Friedman and the author.

Freex and friends in 1973, from left, Bobby Benjamin, Davidson Gigliotti, Todd Benjamin (in stroller), Harriet Benjamin, Carol Vontobel (holding camera), David Cort, Lanesville Postmistress Josephine Devoti, Skip Blumberg, Maria Voorhes of the "Lanesville Country Kitchen" show, Bart Friedman, Nancy Cain (kneeling), Ann Woodward, Sam Ginsberg, Sarah Teasdale and the author. *(Photograph by John Dominis)*

To Carol,
who was there from the beginning

TABLE OF CONTENTS

Chapter		
1	Subject to Change	9
2	Can You Block the Networks?	23
3	Achtung!	37
4	Do They Mix It Up?	46
5	Porcupine Problems	57
6	Something Extounding	68
7	TVTV	82
8	Oh, Yoko!	114
9	What the Clowns Are	122
10	We Don't Mean You!	130
11	Not a Crook, a Spy	140
12	Buckaroo Bart	150
13	Counting Each Dime	162
14	The Lanesville Players	173
15	Newsbuggy	182
16	Elsewhere Area	190
17	You Fallin' Apart Here?	198
	Epilog	209

A schematic of the Lanesville TV pirate broadcasting system drawn by Ann Woodward as part of a larger piece called "Jungletown TV On the Air". The piece was intended for a pricey, coffee-table book on video art but was rejected because the editors did not consider it serious enough.

ACKNOWLEDGMENTS

This is a book of memories, mine mostly, and it would be much more flawed than it is without the generous help of a number of people who shared time and materials that bridged my lapses. Primary among them is Deborah Allen, my publisher, who gave me long-term access to her considerable collection of books on the Catskills in addition to her encouragement and support. Also, there is fellow Videofreex alumnus Bart Friedman, who supplied me with the files from that period and the photo albums he and Nancy Cain, another member of our group, so lovingly assembled. Tom Weinberg opened his files on TVTV, which provided me with considerable insight on that chapter of the group, and Videofreex Bart, Nancy, Ann Woodward, David Cort, Skip Blumberg, Davidson Gigliotti and Curtis Ratcliff, wittingly or not, also jogged my memory in my conversations with them.

I'm grateful for the morning Millard Ruoff spent regaling me with the history of Lanesville in the living room of his home in that community–I only wish I'd been aware of what he knew back when I lived there–and Justine Hommel of the Mountaintop Historical Society graciously shared her rare copy of an unpublished manuscript about life in Lanesville more than a century ago. As innumerable questions popped up, I came to value more than ever the help of Phoenicia Library Director Hilary Gold, and through her, the resources of the Mid-Hudson Library System (MHLS) in Poughkeepsie. I'm indebted to MHLS Director for Development and Resource Management Mary Keelan, who provided me with video books I hadn't been able to locate elsewhere. I also found friendly assistance and early records of Lanesville at the Haines Falls Library. Likewise, the Woodstock Library was quite helpful, as was the staff at the *Catskill Daily Mail*, which let me sift through records in its frigid archives room late one winter afternoon. And I'm grateful for the diligent copy editing of Matina Billias, Patricia H. Davis and Steven Hoare.

Although I spoke only briefly with him for this book, Woodstock historian Alf Evers has been a key source of information through his books as well as a great inspiration to me over the years. The one time I met the late Roland Van Zandt predated the beginning of this project, before I had read his marvelous study of the American Romantic movement, *The Catskill Mountain House* (Black Dome, 1991). But that book was constantly in my mind as I wrote this one. I was also helped by Dierdre Boyle's book, *Subject to Change* (Oxford University Press, 1997), which caught much of the flavor and substance of that time.

My attic still sags from the weight of the Videofreex videotape archive. I frankly admit that I did not feel compelled to view every one of those tapes (again ...) as part of my research. By now some tapes, perhaps many, can no longer be recovered. But the Lanesville TV recordings have been preserved, transferred to VHS format as the result of a free tape copying program of the Sony Corporation. As far as I know, Sony is the only tape manufacturer to recognize the limited lifespan of its product and to offer to clean and copy old tapes without charge. This book would have been far less complete without that service.

I first tried to write the story of Lanesville TV two decades ago, but I was too close to it then, and no one but the Federal Communications Commission expressed much interest in the subject at that time. It wasn't until 15 years later that *Ulster* magazine editor Julie O'Corrozine encouraged me to explore the topic in a long story that led to this book. I'm grateful for the care she took in presenting that piece and for her support and that of Geddy Sveikauskas, publisher of *Ulster* magazine and *Woodstock Times*. Geddy first published my interview with John Glennon in *Woodstock Times* in the early 1980s.

My children, Chloe, Emilia and Sarah constantly provided me with the best incentive an author can have: "Hey, Dad, how's the book coming?" And though she was somewhat wary of my telling this tale–after all, it's her story, too, and I had such an intimate vantage point–my wife, Carol Vontobel, was my best source, ablest critic and staunchest supporter throughout this process. To her, fellow traveler and one of the "Videos," I extend my deepest thanks and love.

P.D.T.
Phoenicia
January 1999

1
Subject to Change

Grayson wore a jacket and tie to work, which immediately aroused my suspicion. But the way he explained what he did, coupled with his manic enthusiasm, made his work sound intriguing, despite the straight outfit. It had something to do with television, but no, don't say television, he said, because that confuses people. Say video.

Grayson Mattingly and his partner, Leon, worked out of an arms factory in a far corner of the District of Columbia. Leon's father owned the factory and let them use some vacant space in back. We walked past stacks of wooden crates the size of coffins filled with oily black rifles, the name of an African country stenciled on the side of each one. Behind the crates, Grayson and Leon had set up a studio where they produced training videotapes for government agencies and private companies. "Aaah know, aaah know," drawled Grayson. "A real turn-aaahf." He was dating my sister, Holly, and I think he wanted to impress me so I'd put in a good word with her.

"Wait'll you see this." He threaded a tape on a machine about the size of my guitar case and played me some of the silly things he'd done: bizarre skits spoofing TV, long, uninterrupted musings, and clowning performed hard against the lens of the camera, all at a pace and in a space unlike anything on TV. No one was telling him what to do and he wasn't selling anything and his antics came across as flawed and foolish and raw, lasting long past the point when I'd decided I must have these same tools because they were going to change television and the world, and I had a part to play in all that.

In 1969, America went to war for a few minutes each weekday evening around 6 p.m. I had dropped out of college a year earlier and hitchhiked to San Francisco. I found the beauty of the West Coast enervating and came back East to await word of my request for conscientious objector status from my draft board in Poughkeepsie. I was bored with psychedelics, disconnected by temperament from the militant core of the anti-war movement, adrift and apprehensive at finding myself with nothing to do in a world where anything seemed possible. And now I'd found something I wanted to do, except I didn't have a clear idea exactly what it was.

With money from a small inheritance left to me by my grandfather on my mother's side, I bought from Grayson all the video equipment I could afford: a used surveillance camera, a 14" TV that doubled as a monitor, a Panasonic videotape recorder (known generically as a VTR) a tripod, some lenses and a device called an inverter for running all this stuff off the battery of a car. Buying the inverter involved a certain amount of foresight: I didn't own a car. So my final video purchase, not from Grayson, was a green and white 1962 Volkswagen bus, into which I built a plywood box to store my equipment. I laid out my blankets on top of that box. I had become an itinerant video artist.

I made tapes of myself and my friends. I made tapes that thankfully have been erased, on purpose or accidentally, of me singing and playing guitar. I made tapes that cannot be played today because I don't know anyone who has the type of VTR I had then. Fate has been kind enough to leave the description of these things to me. My small audiences, people I knew or met as I traveled between Washington, D.C. and Boston crashing with friends, reacted politely to my performances on tape. That's fascinating, they said mildly, as if they'd just watched some minor celebrity they couldn't quite place. Then I would turn on the camera and play back what I'd taped of them and at once, they became enthralled.

Because I owned only one 40-minute tape, I began to build an archive that contained progressively less and less of me and more and more snippets of friends and strangers, a kind of video album that constantly evolved with little design as to where each new entry appeared. These taping sessions made me very popular with my audiences and taught me the first and most important lesson I learned about television: people never tire of watching themselves no matter what they are doing. By extension, any form of video that appeals to the vanity of its viewers is likely to keep their attention.

Another lesson I learned involved how to make the camera an extension of my body as much as my eye. My camera, a small, black, metal rectangle about the size of a thick cigar box, was designed to be mounted behind a bank teller. It had no built-in video screen to serve as a viewfinder. Whenever I shot tapes, I had to watch my TV/monitor over my shoulder, a technique which required some dexterity and which constantly reminded me that more happens in any given space than the camera can ever reveal.

∞∞∞∞∞∞

Early that summer I arrived unannounced at the Washington apartment of a young woman I had met out West the previous winter. Anna opened her door as surprised and happy to see me as I was to find her, and for the next few weeks, we had a delicious affair. It seems odd now to call it that; neither one of us was married or attached in any way, and safe sex meant she used some form of birth control. Only stirrings of commitment could threaten the spirit of free love, and neither Anna nor I felt susceptible.

She accompanied me to the mountainside homestead of a friend in northern Virginia. Outside his 18th-century log farmhouse stood a long-handled pump that had only recently ceased to serve as his source of water. One day, spontaneously as I recall it now, I suggested to Anna that I tape her bathing at the pump. She had long blond hair, a slim model's build, and sweetly put up with my technical insufficiencies. Making love at her apartment, in the bus or any number of other places did not prepare me for this demure video project. As she splashed water on herself, I checked the monitor and it suddenly struck me I had never seen a naked person on TV before. The black and white image of her nudity had such enormous sensual pull it unnerved me. Even as a hippie I retained a prudish streak, and after we viewed this, my second tape, I put the sequence in a box and did not look at it again until a year later, when I discovered technical flaws had rendered it essentially unviewable. I felt a sense of relief.

My other encounter with skin and video happened a few months later in New York City, when I agreed to go with some colleagues to tape a Halloween body-painting party. Body painting involved naked people daubing designs on each other with day-glo colors. I decided to show up a little less than nude, wearing a camera harness strapped to my torso something like a combination crossing-guard belt and flag-carrying brace, with a platform at chin level for the camera. There I stood in the stiff black leather harness, trying to hide behind my camera in the middle of a roomful of naked people holding paint brushes and forcing themselves to behave as if they were having a perfectly natural good time. It's difficult to describe the humbling sensation of trying to function as a cameraman while the people around you slop paint on your genitals.

These experiences convinced me I had no future in the flesh trade. And with the exception of that decidedly unerotic party, I could easily play for my children any surviving tape of mine, including what's left of my earliest efforts, without suffering the least embarrassment over their content. The style of these tapes, however, is another matter.

∞∞∞∞∞∞

Some people take to shooting video with the effortless grace of great athletes who, from an early age, demonstrate innate ability that defies precise explanation. Others remain steadfastly untrainable. I fall somewhere in the middle, having learned my video skills slowly rather than come by them naturally. In addition to this rather fundamental weakness, my hodgepodge of equipment presented me with a number of limitations, not the least of which was that I could only range a few feet from the bus or the nearest outlet because I was connected to the VTR by a coaxial umbilicus. In that sense, my restrictions didn't differ much from the world of conventional television, which confined live broadcasts and taped events to the four walls of a studio except for sports and a few enormously complicated and expensive location specials.

I functioned in a vacuum, too. For all I knew those first few months, no one else in the world made experimental videotapes but Grayson and me, and I seldom saw him. Meanwhile, I ran into dead ends every time I tried to sell my services. I had no shortage of people willing to appear in front of my camera. I just hadn't figured out how to talk anybody into paying for the privilege. I mooched from friends and friends of friends, including Arthur, a painter in Washington, who, perhaps anticipating my arrival, had nothing to eat in his house except a box of sugar cubes. I ate half of them on the sly, rearranging the remaining cubes in the box to disguise how many I'd taken. The sugar cubes comprised all the food and most of the work I had for many days during the summer of 1969.

About this time I saw an ad for the Woodstock Festival in *The New York Times*. News stories, too. The promoters, who called themselves Woodstock Ventures, were having a hell of a time finding a place to hold their rock concert. Bounced successively from Woodstock and then the village of Wallkill, closer to New York City, they settled on a field sixty miles from Woodstock in the Sullivan County town of Bethel, a few miles from where my parents owned a summer cabin. I decided to propose that the promoters include video and me as part of the festivities.

I drove to Woodstock and walked up Mill Hill Road and along Tinker Street. I poked around the hip-looking shops, attracted to them because some combination of beads, music, unfinished wood and incense told me the people inside would know something. Wherever I went, men with beards and women with long, long hair treated my questions wearily, as if they had turned away far more promising seekers. Woodstock's self-conscious attitudes reminded me of Sausalito, California–a trust-funded hipness. But unlike the burn-out I'd encountered so often in the Bay area, Woodstock exuded a sense of anticipation. A place to come back to, I told myself.

I headed south to Woodstock Ventures' sweltering walk-up office one flight above Sixth Avenue in New York. I had invented a company called Video Trips, made up stationery and t-shirts and, as the chief representative of the firm, wore clean jeans and spoke in my most serious voice about the value to the festival of my video workshops, which would introduce people to my vision of video. Fortunately for me, nobody asked for the details of my vision. A tall guy with a beard said all right, sure, fine, go to the site, but stay away from the stage. His conditions didn't bother me in the least. I didn't want anything to do with the stage. The music business had become too commercial for my tastes. Video was new and pure.

I had the official stamp of approval. Tickets cost $17 for each day at this pricey, weekend event, and I could get my friends in for free. Not into the concert itself, maybe, but close enough. Video Trips-about two dozen people from around the Northeast-rendezvoused in the hamlet of White Lake a day before the festival. I arrived late from Virginia and found the intersection by the crumbling movie theater already crawling with longhairs, a comforting sight and my first inkling something strange was going on. Video Trips scouts appeared shortly and led me to the space they'd staked out for us, about as far from the stage as you could get on Yasgur's farm.

The video got off to a slow start. Everyone in our group had already seen my video demonstration, and the people camped around us headed to the music a half mile away, or else settled in for adventures of a chemical sort. Bobby, an extremely short kid I hardly knew, got so high on acid Saturday morning he sat in the center of our circle of vehicles and masturbated until a buddy of his persuaded him to go somewhere more private. Nobody paid much attention and it never occurred to me to tape him.

Dena Crane, a friend who would shortly begin her own video career, came strolling into the encampment a little later wearing nothing but a t-shirt, a reasonably modest costume by that point in the festival. She told me about a booth above the Hog Farm area where a guy had set up a videotape playback system. I was stunned. We all knew the unofficial estimate said half a million people had come to the Woodstock Festival. But someone else with video? I immediately hiked through the swampy woods and up the hillside past the rows of portable toilets to meet him.

I arrived at the booth in the midst of one of many thunderstorms that struck the festival. Clear plastic sheets covered the wooden frame of an oversized lemonade stand, and I could make out the shape of a monitor and of David huddled inside, holding on to the plastic to keep himself and his equipment dry. I crawled in and introduced myself. He asked me did I think any of the monitors might get hit by lightning, and then let out a throaty, hysterical laugh. They're not mine, he said, and if I don't get them back in good shape .... He caught himself mid-thought. But what do I care? he giggled. It's only property, only things.

David Cort had a full head of kinky black hair and a droopy mustache. He wore cut-off shorts and sandal-like, orthotic shoes. He occupied a spot at the mid-section of the festival called Movement City, having set up his booth astride a main route between one of several large campgrounds and the stage. For the last day and a half he had shot videotapes with kids from a radical high school collective in New York City, and he was discouraged. They never show up, and when they do they don't hold the camera steady, he said. They're all over the place. They're soooo stoned. He giggled again in a deep register, an actor's voice.

We talked shop for a while. My inverter had conked out. I couldn't run my equipment. He had power from the generator that supplied the Hog Farm. I'd heard of the Hog Farmers in California, an outgrowth of Ken Kesey's Merry Pranksters, a group that burst into mainstream consciousness in Tom Wolfe's book *The Electric Kool Aid Acid Test*. To me, their encampment defined the epicenter of hipness, and that's where I wanted to be, with video. Just as enticing, David had his own portable, battery-operated VTR and camera, his portapak, the forerunner of today's camcorder. We agreed on the spot that I should bring my equipment to his booth and join forces.

So I lugged my gear to his stand as soon as the rain subsided, though only a few pieces of our two sets of gear proved technically compatible. We talked it over, decided quickly on the superiority of his stuff and used it the next day, Sunday, to roam the site-him interviewing, me behind the camera-taping festival-goers, stoned and straight, Marxist doctors and nurses running the free health clinic, benevolently anarchistic Hog Farmers, a man and his sheep–he claimed they were married–everything but the music, which occasionally crept into the background of the sound track when the helicopters didn't drown it out or the technical flaws of our equipment didn't leave us with long segments shot in complete silence.

David had become pretty woozy by the time he left. One of the high school kids confided in me: Heavy acid, man. I couldn't tell. What did it matter? None of us had gotten much sleep.

∞∞∞∞∞∞

I pulled into Rivington Street in the heart of Manhattan's Lower East Side on a hot evening late in August 1969. It looked as if everyone who lived in the seedy tenements that lined the block had emptied out onto the sidewalk. Jews had left this area years before, replaced by poor Hispanic families. Its east-of-the-Bowery location meant low rents, and a few artists also called this area home. David shared a second floor loft with his lover, an artist named Mary Curtis Ratcliff. Everybody called her Curtis. Downstairs was some sort of business, a bodega, a door away. A musician named Buzzy Linhart and his wife, Jeanie, lived in an upstairs apartment.

I spent that first night in my bus, but I found the noise and the tension unbearable, and I gladly abandoned my streetside accommodations for a small bed next to the floor-to-ceiling windows that made up the front wall of the loft. Directly across the street lived a family with at least six young kids, all of whom lined up at the window each morning with their chins on their hands, grinning broadly, just waiting for me to get up and get dressed. I'd wave and do a little dance to amuse them, avoiding anything too provocative. I felt vulnerable with my bus on the street out front.

Working with the most primitive techniques, a razor blade and splicing tape, we put together a rough edit of the Woodstock tapes, and David wangled an appointment with Don Hewitt, the producer of a new CBS news magazine show called *60 Minutes*. Hewitt eagerly watched the footage, then dismissed it as old news and proceeded to lecture us, settling on the example of John Lindsay. He predicted Lindsay could never be re-elected mayor of New York because the country had turned conservative, and people like us would quickly become conventional. He looked at David. "How old are you?" he demanded. "Twenty-five? Twenty-six?"

"I'm thirty-five." David was wheezing his nervous laugh.

Hewitt jerked his head back. "You are not!" he shouted. He sounded shocked. I was surprised, too. I hadn't asked David how old he was, although I knew he had several years on me, but fourteen was more than I expected. It didn't matter to me, but it did to Hewitt, who went on and on massaging his disbelief, as if he couldn't accept someone of David's age and obvious education would choose to look–and live–the way he did. Hewitt didn't give me a second glance. Just another hippie.

Reporter Mike Wallace stopped us on the way out and asked what we were up to. "You just had more time with him than I've had in the last month!" said Wallace, pointing at Hewitt's closed office door. We shrugged, unimpressed, and strolled out feeling cocky, thinking we didn't need the straights at CBS.

One thing we could have used from Hewitt was a place to park. David had cavalierly pulled his dilapidated Rambler up on the curb behind CBS and the cops had towed it away.

∞∞∞∞∞∞∞∞

A month or so after that meeting, we returned to the CBS Broadcast Center on West 57th Street, this time for a kind of private garage sale. A kid who worked in the CBS mailroom as a summer job had met David at the Woodstock Festival and subsequently described the encounter to an executive at CBS named Don West. Don had started out in radio in New Mexico then moved east to Washington and up the broadcast industry ladder as a print journalist, eventually landing a job as special assistant to CBS President Frank Stanton. Don desperately wanted to produce his own TV show. He couldn't bear to see the '60s pass him by, and when he came to the loft and David handed him the little Sony portable, it was the first time in a career in television Don had ever held a TV camera or shot a tape. He got hooked on us, and we got hooked on his promises, never stated explicitly but frequently intimated, that we could do our thing on the CBS Sunday night prime-time slot made vacant by the network's decision to can the Smothers Brothers.

Never mind that the network had decided to cancel Tom and Dick Smothers' show because their mild brand of humor had frightened CBS executives into believing the program had become too controversial. We had no reason to doubt Don. He could arrange for us to wander the bowels of CBS and pick out any used equipment we wanted, even if we did reject most of it as antique junk. Don offered us a ticket to a big time audience. We need only do our thing, which amounted to whatever we felt like doing each day; never mind pursuing a specific concept for a show. This didn't seem disorganized or vague, just cool, mellow and supremely hip–the very qualities Don had in mind.

We taped all over New York that fall: political theater, music and dance, the body-painting party. On Don's money, much of it directly from CBS, we went to Chicago for the trial of the Chicago Seven (the Chicago Eight before the infamous Judge Julius Hoffman had Black Panther leader Bobby Seale brutally and illegally bound and gagged and removed from the courtroom). We interviewed Abbie Hoffman (no relation to Julius) and Jerry Rubin of the Yippies, and SDS leader Tom Hayden, all of them defendants in the trial. But when we told Hayden of our involvement with CBS, he insisted we erase all tapes of him. He was certain the tapes would end up in the possession of the network, which would allow authorities to use them against him. We scorned his concern as paranoia, asserting the tapes belonged to us. We would protect them and him. But he was adamant, and we agreed to his demand because we felt obligated to prove our radical bona fides even at the expense of the tapes. As it turned out, he was right. CBS did get our tapes. He wasn't on them.

On that trip we also taped Fred Hampton, chairman of the Chicago Chapter of the Black Panther Party. Less than two months later, Chicago police murdered Hampton. We had to repossess the Hampton tapes from CBS late one night, hidden in a guitar case.

∞∞∞∞∞∞∞

No happening show in 1969 could overlook the high life on the West Coast. So, late that fall, shortly after Lindsay won re-election despite Hewitt's dire predictions, we packed up all our equipment and flew to California. In Los Angeles, we hopped off the plane into a huge, rented RV and trucked around the state for a couple of weeks in a kind of whacked-out version of *On the Road with Charles Kuralt*.

When we returned, we found Don had rented us a place in the Greene County hamlet of East Durham, which looked out on the imposing eastern escarpment of the Catskills known as the Great Wall of Manitou. Snow fell early and heavy that year, forcing us to pay a little more attention to our surroundings than we might otherwise have done. Most days, though, we shut ourselves in an upstairs bedroom, pulling the shades and focusing all our energy on the monitors. East Durham gave us our first collective taste of the Catskills, and it may well have affected us more than we knew at the time. In this majestic setting, with all we wanted to eat and drink, we edited our tapes free of the distractions of the city. We saw ourselves as rock stars now, and this show for Don, our first album. We planned to be big. Very big. We had two weeks and no margin for error.

Along the way we had acquired a name: the Videofreex. We thought of ourselves as a group, though officially we had incorporated as a business. Besides David, Curtis and me, there was Davidson Gigliotti, a master carpenter and the scion of a wealthy Connecticut factory owner, who had accosted David on the street one day that September to ask about his video equipment. Davidson wore crusty jeans, scuffed boots and a stained t-shirt ripped on one sleeve. His front teeth were in sorry shape, too. He was scruffy even by my low standards. He followed David back to the loft; we talked for a while and David and I figured we'd never see him again. But a few days later, Davidson, who'd studied at the Rhode Island School of Design and once worked for UPI as a sportswriter, showed up with a video system of his own. A few weeks later, he volunteered to build the boxes to house our equipment on the RV in California. We didn't know at that point—couldn't have from his exterior—that he was a perfectionist. He had trouble finishing things on time. He drove me crazy putting the last crate together as the airline freight handlers headed up the elevator to the loft. He was a counterweight to the rest of us, who had reconciled ourselves to the imperfections of the young medium.

And there was Chuck Kennedy, a chain-smoking, beer drinking, compulsively inventive bench technician for a video equipment dealer. Chuck reacted warily to David's offer. He had a good job. Why should he risk it? We sweet-talked him into joining us but just, we agreed, for the duration of the CBS gig.

∞∞∞∞∞∞

The name Videofreex played off a term about to become generic: video freaks. Among the hip, the pejorative sense of freak had been subsumed by the counterculture notion that any enthusiast, regardless of the subject—sex, drugs, rock 'n' roll, food, video—qualified non-judgmentally as a freak. Video freaks encompassed all those artists and activists who had begun to experiment with low-cost videotape recording equipment similar to mine and David's.

Sony introduced VTRs using half-inch-wide tape on open reels, the ancestors of today's ubiquitous VCR, as educational tools and consumer toys in the mid-1960s. In 1968, the company began marketing a relatively small, battery-operated VTR with a lightweight, black and white camera, the system David brought to Woodstock. This rudimentary technology forced us to improvise all the time, if we wanted to mimic the simplest broadcast TV techniques. Our attempts frequently failed, and plenty of long taping sessions ended up as nothing but salt and pepper on the screen accompanied by the gnashing of teeth and wails of dismay from video freaks who felt betrayed by their tools. We persevered because we believed that however technically unrefined our equipment, these small machines of ours held revolutionary promise.

Until half-inch VTRs came along, the means to produce television programs had resided almost exclusively with the networks because only they had the money, equipment and technical support needed to make TV shows. Now, suddenly, the tools of television production had fallen into the hands of people not beholden to the networks in any way. Video technology fit neatly with the revolutionary ethic of the time in that it didn't matter so much what you produced so long as you didn't do what they–the broadcasters–did.

Our connection with CBS left Videofreex not quite so politically pure as the revolutionary rhetoric we embraced. But almost no one in our circles other than Hayden had raised the issue of the potential contradiction between what we were doing and who we were consorting with to get the tools to do it. We took that as an indication no one cared to dwell on the fine points of ideological pedigree. We certainly didn't. And so what if our provenance was politically blurry? Here we were, ready to assume our rightful role among the elite, if not the vanguard, of the counter-culture and anti-war movements. With our contemporaries, this gloss-over seemed to work. Our network benefactor was another matter.

∞∞∞∞∞∞∞

Don showed up at the East Durham house for a weekend to look at our footage, but we had just started editing, and neither we nor he could grasp what we had or where we were headed with it. His visit ended with his own wistful admission that he didn't fit in no matter how hard he tried. He left after two days, though he could easily have insisted on staying. A few days later, a serious, short-haired guy named George Pepper arrived, saying Don had asked him to drop by. Pepper lived in a Boston commune run by a messianic character named Mel Lyman. Deserved or not, Lyman had been accused in the alternative press of using strong-arm tactics as a proselytizing technique. Don liked Pepper, thought his watchful manner signaled that he was a very "powerful" guy. We viewed him as creepy, a spy for Don, and weren't reassured when he hinted he was wearing a pistol under his jacket. Pepper lurked for a day and then announced he had to leave for Manhattan to supervise Davidson's construction of a TV studio for the presentation to CBS.

We finished editing and returned to the city less than 24 hours before the presentation was scheduled to go on. At our new loft, we found something like the outlines of a control room. Pepper greeted us, scowling. He was covered with sawdust and he wasn't wearing a gun. Behind him, Davidson calmly banged away at the studs to frame a control room, working at his own, perfectionist pace. He showed no sign of concern over Pepper or the looming deadline. This state of affairs seemed acceptable to us. What did we know? We'd never produced a live TV show before. Why worry now?

Don had even less production experience than we, but he did have a sense of the price he'd pay for failure, and he was having some second thoughts.

Fred Hampton had been murdered by the Chicago police just a few days earlier, and Don had gotten ahold of news footage of Hampton's funeral. He thought it would lend credibility and timeliness to the show, and he wanted to lead into our taped interview with Hampton with scenes from the funeral accompanied by a dirge. I was incensed.

It wasn't so much the funeral. That might have worked. The problem lay in the nature of the funeral footage: It was film. Film was the old medium, distant, inflexible, elitist, impure. And who was he to interfere with our production? I tried to reason with him as we stood alone in the control room less than an hour before the show. He wouldn't budge. I didn't see that I had a choice. "Don, if you try to play that stuff about Hampton's funeral, I'll disconnect so many cables you'll never get this show on."

He stared at me, then down at his feet. All around us was a rats' nest of wires running every which way. Very softly he said, "If you mean that, we'll never work together again."

I stood there, head cocked to one side, until he'd left the control room. I didn't have time to figure whether or not I'd really won something.

∞∞∞∞∞∞∞

The man in charge of CBS TV programming in 1969 was Mike Dann. He had helped bring *The Beverly Hillbillies* to the network as well as a host of other shows best left unremembered. His assistant, the head of daytime programming at CBS, was Fred Silverman, who went on in his career to head–some would say to ruin–programming at all three major networks. On December 17, 1969, a bitterly cold night when the city already had several inches of snow on the ground, Dann, Silverman and a third CBS executive, Irwin Siegelstein, traveled by limousine to a warehouse district in lower Manhattan to view our live show on closed-circuit TV. They came not because of Videofreex but because the show had been produced by Don, the assistant to their boss, and because CBS had bankrolled Don and they wanted to know what they'd paid for.

Immediately after the show, Dann, who seemed a little drunk, came out of the adjoining loft where he, Silverman and Siegelstein had viewed the presentation on a monitor. He spoke to all of us who had worked for several months to put together *Subject to Change*, Don's name for a program which, scattered though it looked, contained elements strikingly different from anything then on television. Condescendingly, Dann referred to the show as an "experiment," which, in fact, it was. He said a show like *Subject to Change*, a mix of documentary and live music–Hampton, Buzzy Linhart, an alternative school in California, a bluesman named Major Wiley, and so on–segments not tied together with a host or omniscient narrator, was at least five years away from appearing on network TV.

In one way, Dann's prediction proved accurate to the year. In 1974, TVTV, a group of independent producers, including some members of Videofreex, put together a quirky documentary on a 15-year-old guru that aired over the relatively new Public

Broadcasting System. Though PBS had come into being to offer a non-commercial alternative to the networks, it, too, resisted unconventional ideas. The guru tape proved the exception, not the rule, as far as independent productions were concerned. Creative work had to look elsewhere for its audience. But neither Dann nor we could have known any of this that night in 1969. He had delivered his unappealable verdict on our presentation and we were not really surprised. He and his colleagues didn't get it because they and the small audience of invited studio guests had witnessed a show that didn't aspire to the standards of commercial television, or even to those of non-commercial TV. What we had produced amounted to anti-commercial television. At that point, anyway, we could find no bridge between Dann and us long-haired, scraggly, anarchistic hippies, because we weren't in it for the money. It didn't bother us in the least that we couldn't define with unanimity what we were in it for or how we expected to pay the bills.

∞∞∞∞∞∞∞

Don West disbanded his small production company after a vain attempt to re-edit the show to please Dann. Dann belittled Don's effort as "a piece of shit," and the network told Don to clean out his desk. There was no place at CBS for the people he'd hired, and two of them, Carol Vontobel–she and I had become a couple in East Durham–and Nancy Cain, joined our group. In a very short time, Don's project had allowed us to gain an immense amount of experience and a formidable array of equipment, far more advanced than anything anyone else working independently in video in Manhattan or elsewhere around the country could brag of. And we had something else: We had an identity. We were the Videofreex.

We saw ourselves as number one on some imaginary video chart, though that didn't mean the other video groups in New York City and around the country necessarily agreed with our assessment. The city alone nurtured half a dozen thriving video groups, with names like Raindance, People's Video Theater and Global Village, all of which, at various times, we collaborated and squabbled with.

At the outset of our CBS project, we had set up our studio in a fifth floor loft in an industrial district in lower Manhattan that shortly became known as SoHo. Desolate on weekends and at night, SoHo on workdays offered an unending parade of trucks that whammed, banged and honked their way down narrow, cobblestone streets. They created a din at sidewalk level that exceeded the pain threshold and which, even after it wafted up to the fifth floor, still defied our best efforts to filter it out of our recordings. We held regular Friday night showings of our tapes of politics, artists and rock concerts, happenings around town and flights of fancy. To attend meant climbing five long flights of stairs in near darkness. Davidson volunteered to stand by the door next to his upturned Stetson, glowering at anyone who overlooked our sign requesting donations. His brooding presence boosted the contributions substantially, but it didn't pay the rent.

Carol and Nancy got real jobs. Chuck couldn't face going back to punching a time clock and fixing other people's machines, so he quit his job, gave up his penthouse apartment, and came to live in the loft, where he was joined later by Ann Woodward, a young woman he'd met at a video art show at Brandeis University. Carol's friend and fellow elementary school teacher, Skip Blumberg, joined, too, as did Bart Friedman, a sometime talent agent, record producer and cab driver. Various others came and went.

∞∞∞∞∞∞

For all the excitement about video among freaks and the straight press, it might well have become a short-lived phenomenon, and Videofreex with it, if not for the contribution of one man who had no noticeable interest in the medium and who never looked completely comfortable on TV in any of his frequent appearances. What he cared about was his own legacy, and he had very definite ideas about what that legacy should be. His name was Nelson Aldrich Rockefeller.

In 1970, Rockefeller's chances for the presidency had all but vanished. Richard Nixon, busy extending the Vietnam ground war into Cambodia just as the Army brought officers to trial for a massacre at a small village called My Lai, would undoubtedly win renomination by the Republicans in 1972. That would leave Rockefeller, in that pre-Ronald Reagan era, too old to run in 1976. So the four-term governor of New York sought to leave his imprint in different ways. For one thing, he had decided to build a grand plaza of government office buildings in the center of Albany, a kind of titanic monument to his tenure. But Rockefeller was too sophisticated to believe he could secure a lasting memorial to his accomplishments simply by altering the architecture of the hinterlands. So he took another bold step, and in the guns and butter economy of that time, he reinvented a small Bureaucracy he had created years earlier and made it into a major state agency with far-reaching cultural power. The New York State Council on the Arts budget grew from a total of $2.2 million in 1969 to $20.2 million in 1970. The 1970 Council budget earmarked over a million-and-a-half dollars for the Film, Literature and TV/Media Program, an unheard-of sum for a government to dispense on the arts in those days, particularly when part of the money was earmarked for a new medium called "media."

Rockefeller resisted the philistine impulse to meddle, and that allowed the Council to take the radical step of including artists themselves in the process of determining what and who should be funded. The Council appointed panels of artists in each field, with each panel asked to make specific recommendations for how to dispense its share of this new largess. For the panel deciding on Film, Literature and TV/Media Program grants, it was clear that writers and poets didn't face the overhead that burdened film makers and video artists, and the bulk of the money would have to go to the makers of moving images. The partisans of spending heavily on film dominated the ensuing discussions because film was the established medium; and the needs of filmmakers might have swamped the upstart video movement altogether except that for all kinds of artists from practically every discipline, the new, easy-to-use equipment had made video a medium too intriguing to ignore.

The governor also made some astute calculations to assure passage of his plan. New York City might claim its place as the hub of the cultural universe, but if all the money went there, he couldn't win the support of upstate legislators, many of whom harbored misgivings about the notion of subsidizing the arts with taxpayers' money. So the word went out that the Council would spend a sizeable chunk of its funds throughout that netherworld known as "upstate."

David got tipped off about this policy, and we immediately set out to play the game to our advantage. Videofreex officially transmogrified into a non-profit organization called Media Bus. Originally, we intended to equip a bus with video equipment and travel all over the state. The idea owed a lot to a New York film maker and film teacher named Rodger Larson, who had developed the Film Bus with Council funds a few years before and had traveled to neighborhoods around the city showing all kinds of movies. The concept of a bus of any kind had a certain mystique, too, because of Kesey and his adventures. But before long we concluded that a heavily-equipped bus mimicked the networks. Portapaks, our portable VTRs, symbolized the free-spirited approach to video we planned to promote. So we ditched the bus concept in favor of a van. Stay nimble, we said, ready to shoot tape at a moment's notice.

We worked it out that money for Media Bus would come through the Rochester Museum and Science Center. All $40,000 of it. After months on end when it looked as if the group could not stay together for lack of funds, we could hardly imagine how to spend that much money in a year. Even as the announcement of the grant arrived, Con Ed cut off our electricity at the loft and we could only work now and then by running an extension cord to the neighbors below, hoping we wouldn't blow their fuse. Several months behind on the rent, we had to flee the city.

∞∞∞∞∞∞∞

No one in the group knew anything about Rochester except that it lay some distance north of the Bronx. But we didn't have to live in Rochester to get the money. We might have broken up right then, but we couldn't sort out one person's equipment and tapes from another's. So, by default, we decided to depart as a group, though not without some serious misgivings.

Our first house-hunting probes struck out in Sullivan County. Stockbridge, Massachusetts, which seemed to us close enough to the New York border to qualify as "upstate," proved too upscale. Woodstock looked beyond our means as well. Real estate agents everywhere acted unsure of what we wanted or else sure they had no intention of inflicting the likes of us on their clients, especially in light of our preposterous claims to need space for a TV studio.

Carol, Nancy, Bart and I, the desperation house hunting crew, were on our way to Woodstock from Stockbridge, tired and discouraged, when we passed through Haines Falls, which sits atop the Great Wall of Manitou. I remembered the name from ski trips to nearby Hunter Mountain. I always thought of the small Catskills settlements along the way to the slopes as bleak and tacky, and nothing I saw as we stopped there changed that impression. It was early spring and the ski season had ended. The leaves had begun to emerge in the lowlands, but not up here. The grass lay flat and brown, tree limbs barren.

Louise Doyle, an agent we picked at random because her office was alongside the highway said, yes, she had some big homes that could accommodate a group. She didn't act put off by our scruffiness. We followed her car for miles, through the village of Tannersville, scrubbed gray by the winter and all but abandoned at that interim time of year, then left onto a smaller road, Route 214, and through the notch, where the boulders from old landslides crowded the shoulders of the road. After that, the forest closed in, broken by clearings for trailers and clapboard houses as we passed Edgewood headed down to Lanesville.

Louise showed us a house of some elderly folks at the end of a short cul-de-sac called Neal Road. It wouldn't have fit all of us and the equipment. Another disappointment. She insisted she had more places for us to see. I looked out the window and across the side lawn at the huge house next door. We needed something that big, but I didn't say anything. A week later Louise called back. She had a listing for the big house I'd seen. Within a month we'd rented it, signing a one-year lease on Maple Tree Farm starting in June 1971. As quietly as we could, we packed up everything in the loft on Prince Street and at our West End Avenue apartment and snuck out of the city for Lanesville.

2 Can You Block the Networks?

"I eat possum an' I eat squirrel, but di'n't eat yer dog."

Gene was shaking my hand, but he didn't seem to notice and wouldn't let go. He had a funny way of breathing through his nose that sounded like an air brake on an 18-wheeler. He had finished talking. There was only the sound of his breathing. My arm hurt.

Until that moment it had not occurred to me that anyone or anything had eaten Mushroom. Now, confronted with his denial, I was at a loss how to respond. Did this mean he had eaten the dog and wanted me to absolve him? It was chilly outside and he wore a t-shirt. I invited him in.

"You got any vo'ka?"

"Just beer," I said. He must have known it was a brand called Munich, which the new owner at the Lanesville General Store dubbed "hippie beer," and which we disdained as "camel piss" but drank anyway because it was all we could afford. Calling it "hippie beer" gave Denny a way of expressing his disapproval of us while taking our money. He had to do business with the Videofreex. The ten of us comprised a significant portion of Lanesville's population.

Gene Grant, a squat, moonfaced chunk of Catskills bluestone, was no hippie, and as far as I knew, he didn't disapprove of us. Only our taste in alcohol. He shook his head and turned off into the darkness. As an afterthought, he let go of my hand. That was it. He never said another word to me about Mushroom.

∞∞∞∞∞∞∞

I first encountered Gene in June 1971, the day after we moved to Maple Tree Farm. We were out early, enjoying the mountain scenery, reminding ourselves we were no longer in the city, when Gert Neal pulled into the driveway in her garbage truck and, seeing her exit was blocked, drove across the front lawn and down Neal Road. Gene was sitting next to her in the cab and they waved as they went. We figured it was a local custom. Like not eating your neighbors' dog.

We wouldn't let a single rusty, mufflerless, half-ton dump truck tooling through the yard bring us down. Whatever our individual apprehensions, the bright sunshine playing on our new green world overcame them. We had almost given up trying to find a place for all of us by the time we stumbled on Lanesville and Maple Tree Farm. And now that we had made our decision to move here, we intended to take full advantage of our new situation. To deliberately leave New York City, a place where we could almost always round up an audience to watch our videotapes, where we could rail against the war and tape others even more committed than we, where our studio could host the most outrageous events without anyone so much as blinking because city people expected no less, and where we could move in circles of like-minded artists–to leave New York behind for who knew what, here in a place no one had ever heard of and which didn't appear on most maps, to come here alone except for each other, ranked as a lot scarier and more unsettling than any oddball reception by the neighbors.

∞∞∞∞∞∞

Even then, in the midst of a minor renaissance, you'd be hard-pressed to call Lanesville a hamlet. It lies along the only near-flat stretch of State Route 214 as it rushes north and east up the Stony Clove Valley from Phoenicia to Tannersville and Hunter. The rump of Hunter Mountain, where it forms one side of a natural bowl leading to a geological cleft known as Diamond Notch, dominates the landscape. The top slope of Hunter, at 4,040 feet, is the second highest point in the deeply riven northeast corner of the Allegheny Plateau known as the Catskill Mountains. And Lanesville sits at the heart of the Catskills.

The Stony Clove Creek flows down through Lanesville emptying into the Esopus, which in turn fills the Ashokan Reservoir with water for the plumbing of New York City. And like the creek, the people who live along this valley tend to gravitate downhill across the Ulster County line to Phoenicia, and downhill again to the Hudson at Kingston rather than uphill through the precipitous notch above Devil's Tombstone and over the mountain to Hunter. Although Lanesville lies within the boundaries of Hunter township in Greene County, the villages of Hunter and Tannersville, the main settlements, lie within a different area code. Snow can make them hard to reach in winter, and back then, at least, the town provided sparse services. Longtime residents claimed they never saw town officials or representatives from the county seat in the Village of Catskill except at election time, when they didn't offer politicians a cordial reception. But you found jobs in Hunter at the ski slope and all the spin-off businesses, and the school, too. And the tug between the areas gave Lanesville two of its many personalities.

Maybe perceptive historians could read Lanesville's past from its external appearance. We couldn't. We knew it only as a place where we had finally found a structure big enough and cheap enough to house us and all our equipment. We told each other our isolation did not amount to exile, what with the city less than three hours away by car. And like so many generations before us, we were happy to have escaped the grit and din and hassle of New York. Most of us, anyway. We were so focused at first on getting settled that we took little notice of our immediate surroundings other than to remark to each other that this was really out in the country.

Lanesville had half a dozen old boarding houses like Maple Tree Farm. They were universally clapboard-sided, almost all of them white. Many had been extended and expanded piecemeal over the years, disguising original lines with a bewildering array of wings, dormers, annexes, turrets and, at the Farm, a bridge of sorts–actually a second story hallway that connected the main house to a smaller structure in the rear, with the open driveway below. These buildings exuded a sense of haphazardness, of inconsistent craft applied hastily in response to unexpected booms in the tourist trade, and an unfinished quality that reflected times when everything suddenly went bust.

Reconstruction of Route 214 after World War II had rerouted sections of the road and cut a deep gash in the front yard of Maple Tree Farm. The new road altered the landscape and the relationship of the structures on it in that the wider, straighter road made it easier to speed by and never give Lanesville a second glance. Downtown Lanesville consisted of a gas station and grocery store all in one, called the "general store." That building, which also housed the post office and a family-sized apartment on the second floor, wore fake brick asbestos shingle siding. Its gas pumps lay so close to the new road you felt as if you took life in hand to step out of the car on the highway side.

Across the street stood an abandoned store with large plate glass windows framed by a porch, the general store before the gas station/grocery came along. On one side of the empty store sat a few modest, two-story frame houses, and then a rambling boarding house badly in need of a coat of paint. On the other side, a small junkyard. To the south, just above the county line, Beecher Smith's sawmill, the only industry in the settlement, guarded the entrance to the upper valley. Doyle's Bar, a long, low structure that looked a little like a chicken coop, sat on a vestigial section of the old Route 214 roadbed between the sawmill and the hamlet center. The rest of the community spread out in a mix of homes–a few neat, suburban-style houses rose alongside aging trailers, some on real foundations, some on blocks, with porches or small additions, and only a few with the wheels still showing, let alone useful. In the junkyard, an ancient shell of a trailer listed so far to the side you couldn't tell what kept it from toppling over.

The homes of Lanesville lie along the highway or on one of the half dozen or so side streets off 214. With the exception of the flats behind and south of the general store and the road leading to Diamond Notch just west of the hamlet center, no one could live very far from the highway. The mountains close in too abruptly. At Maple Tree Farm, a quarter mile up the road from the general store, Hunter Mountain swept steeply upward just outside our back door.

The last glaciers more than 20,000 years ago had seen to that. The mile high ocean of ice that had edged inexorably south from Labrador had bumped up against Hunter and Plateau mountains and the rest of the Catskills' Central Escarpment and then spilled through the notch and down the valley. As the continental ice shelf contracted to nothing, it deposited its collected treasure down the backs of these ancient mountains, narrowing the valley with lumps and folds of stone, and precious little soil.

On July 22, 1971, four weeks after we moved to Lanesville, the Pentagon proudly announced eleven American servicemen had died in Vietnam the previous week, the lowest number of U.S. deaths in a week since 1965. A quarter million U.S. troops remained in Southeast Asia that July, fewer than half the number of men at the peak of our military build-up. The administration saw this latest body count as a sure sign the President's program of turning the war over to the Vietnamese was working. As if to underscore the view that Washington finally had Vietnam in hand, the man our landlord, Sam Ginsberg, called "Nixon, de sonofabitch bum" had announced a week earlier that he would visit China, becoming the first American president ever to do so. Nixon wanted a legacy grander than losing the ugly little war he'd inherited and extended. His trip to China would reposition American foreign policy and distract the nation from the war. Some of the nation, anyway.

At Maple Tree Farm, we followed these developments in the *New York Post*, a liberal tabloid and the only mass media outlet that publicly expressed mistrust of Nixon's Vietnam policy. A month earlier, *The New York Times* had printed the Pentagon Papers, which detailed the official deceptions that had led to our full-scale involvement in the Vietnam war. For us, the Pentagon Papers was old news; the anti-war movement, through its contacts in the European press and in North Vietnam, knew U.S. media were regurgitating government misinformation about our role in Southeast Asia. The decision of the *Times* to run the Pentagon Papers despite the ire of the government may have been remarkable from the standpoint of corporate media, but it struck us as too little too late. The monster of the war machine lurched ahead balancing the body count equation by upping the slaughter on the other side. And the tabloid was the only mainstream source of outside information available on a daily basis that dared whisper of that.

In the city, exposure to news is an inescapable fact of life. The headlines on the newsstand enforce a threshold awareness. In shop windows, restaurants, cabs, buses and subways, the papers, radio and TV insist we absorb the daily minimum dose of information. Yet Videofreex, despite our chosen medium and our city origins, had no particular commitment to reporting news. Even if we had, how would we have delivered it–and to whom? We had burned our bridges with CBS, and no other network would have touched troublemakers like us. New York City was still in the process of stringing cable for cable TV. And public television was in its infancy, as its leaders industriously fortified their system against any temptation to experiment.

We tried to peddle our tapes around the country to colleges and like-minded, media-savvy individuals and groups with VTRs compatible with ours. We found few takers. On Friday nights, we had offered anyone willing to brave the five dark flights of stairs leading to the Prince Street loft the chance to watch our latest tapes. For a particularly hot topic, say, the Hells Angels motorcycle gang, we could almost fill the meager seating in the studio; more often, the loft was nearly deserted. The novelty of these shows wore thin.

We had failed to develop a loyal audience willing to make that arduous climb for uneven presentations that often verged on technical calamity.

Not that any of these obstacles shook our faith in the power and future of video. We had made a commitment to work around the distribution roadblocks on our own terms, whatever it took. We were convinced video wasn't a gimmick, it was a tool. But a tool for what? We had rejected the rules of authoritative, aloof, jacket-and-tie reporting that dismissed anyone who had anything uncomfortable to say. And that decision had cut us off from any chance to reach the mass audiences of broadcast television. We clung to the faith that our art, our independent video, our alternative to the media, would change public consciousness if only we could get it to the public.

So here we had landed in Lanesville, which confronted us with an abrupt disconnect from the sources of information we took for granted. The general store carried the *Post*, and if one of us, most often Skip, got there at just the right time that first summer, we'd still find a copy left on the counter. If we didn't, we went without news for the day. We seldom listened to the radio except in the car, partly because there was no reliable reception until you got well clear of the Stony Clove valley. The nearest daily newspapers came from Kingston and Catskill. Both had conservative slants, cared little about the world beyond their municipal borders, and nothing about Lanesville. Not surprising. Nothing much happened here that would qualify as daily news. That left television, which the mountains all around us blocked out, as far as we knew.

When we'd first come to Maple Tree Farm to sign the lease with Sam and Miriam Ginsberg, their old black and white console TV sat in the middle of what became our music room. The images on it were indecipherable–ghosts in a snowstorm–although I could hear the sound well enough to make out an *I Love Lucy* rerun. "Sometimes it gets better when you monkey mit de antenna," said Sam. He was delighted when I offered to try, and after I'd spent a few minutes twisting the flimsy mast that held his antenna six feet above the lawn, he told me I'd improved the reception. The picture looked like the same fuzzy nothing to me.

We moved in a few weeks later and discovered Sam had taken the antenna with him to his new house.

∞∞∞∞∞∞∞

Though our arrival had coincided with a lull in ground combat in Vietnam, the war hovered over our lives like an unwelcome visitor whose sudden appearances we could never predict. Of the six men in the group, only Chuck had been in the military, and he had been mustered out of the Army with a general discharge after a court martial for an incident involving a wrecked jeep. He very likely would have been shipped to Vietnam for a tour of duty had he not departed the service ahead of schedule. The rest of us had escaped the draft for various reasons. I had marched against the war, a body in the crowd, not a leader. I had toyed with facing prison for my pacifist beliefs, but my draft board obliged my request for conscientious objector status. Most other Videofreex had had glancing involvement with the antiwar movement. And though none of us was an ideologue, all of us, the men and the women, shared an unshakable opposition to the role the U.S. played in the war, having arrived at our views independently well before Videofreex.

No one in Lanesville openly opposed the war. But the war had come there. Betty and Scotty Sickler down the road had lost a son in Vietnam.

The anti-war movement of the late 1960s and early 1970s had become so intertwined with counterculture as to make the two movements inextricable. We considered ourselves part of both. We would tape an artist's happening one day and a protest against the war the next. And no one embodied the spirit of the two movements better than Abbie Hoffman. He had gone to the South as a civil rights activist. He was a pied piper for runaway youth in the East Village drug scene. He appointed himself the ringleader of the Yippies' efforts to disrupt the '68 Democratic Convention in Chicago, as well as other successful political theatrics, including the tossing of dollar bills onto the floor of the New York Stock Exchange, which halted trading as brokers scrambled for the cash. At the Chicago Seven conspiracy trial we had learned that when Abbie showed up, his gleeful spontaneity and comic insight would yield an amusing tape with political punch.

Before we left the city for Lanesville, I received an invitation to join a group of people, including Abbie, going to Cuba to meet with young people from North Vietnam. We were to travel with the Venceremos Brigade, a group of U.S. radicals who, in defiance of the State Department ban on travel to the island, sailed to Cuba each year to help with the sugarcane harvest. Our group was an ad hoc gathering of artist-activists, most of us in our twenties or thirties. It was never clear to me precisely how I was chosen, but I was flattered and excited. A series of meetings followed to plan for our visit. There were tensions from the beginning, especially about the self-appointed leadership group, with some of the feminists pointing out that the leaders were all men.

One meeting was held at our Prince Street loft. Everyone arrived at about the same time, out of breath after trudging up the stairs. The moment Abbie entered the room he began juggling. He did it naturally, as if he always made his entrance that way. It immediately drew attention to him and away from the group of twenty people with him. Everyone now knew Abbie had arrived. A few minutes later as we sat in a circle in the studio, a woman questioned the travel plans, nothing political, just a practical point. Abbie pointed across the circle at her, "You don't like me and you've never liked me." It was not a complaint; it was a challenge.

I had gotten used to harsh words in the sporadic group meetings of Videofreex, but I'd never witnessed such an overt play for power. I often attempted to mediate at our meetings, but here I sat as startled as everyone else while Abbie took control. Perhaps he wanted these issues resolved before we left for Cuba. But Abbie's actions left me uneasy about the standards of the counterculture and whether we were evolving a higher consciousness after all.

The trip to Cuba never happened. We were told peace negotiations in Paris had reached a critical point and our visit might upset the diplomatic balance. Each member of our group received a postcard from North Vietnam of Ho Chi Minh. The spelling of my name on my card was almost unrecognizable. I wrote it off to the language barrier, but always wondered whether they did it intentionally to protect me from the forces here that already had my name. Perhaps the Vietnamese people, at war for a generation, recognized the value of such things as deniability. Onto the card, they had taped a ring, forged from the metal of an American warplane shot down over the North. I took pride in the card, but the ring struck me as a ghoulish souvenir no matter how just the cause. I wore it once, uneasily, then put it away.

Not long after the trip was canceled, Abbie asked me to meet with him. He said he had a contract to write a new book and he wondered whether I would contribute to it. He wanted a set of instructions on how to set up a pirate TV station. I didn't know the first thing about broadcasting at that time, but I relished the assignment and immediately agreed to give it a try. The title of his manuscript was *Steal This Book*, so I went to a university library and stole a paperback book on the physics of television. (I still suffer remorse for not returning that paperback, which, fittingly enough, someone borrowed from me and never returned.)

I delivered my manuscript to Abbie at the Bowery loft of poet John Giorno. I still wasn't sure exactly what he wanted this for. My research seemed abstract to me, so I pressed him to tell me what he had in mind. Abbie had a true showman's grasp of the value of the absurd. He wanted to black out TV transmissions in New York City and insert programming of his own in its place. He became more animated as he warmed to the topic. He envisioned this guerilla action, a kind of media hit-and-run. The transmitter would have to be small enough to escape detection so he could use it repeatedly. He grinned but he was serious. Could we put it all in a Volkswagen bus?

I couldn't answer for sure, but the breathtaking scope of the plot enthralled me. Both of us had seen too many World War II Hollywood movies in which the underground heroes use their clandestine transmitters to outwit the Gestapo. I couldn't let this opportunity go, regardless of the consequences. And I owned just the type of bus he had in mind. But because I doubted my academic research on the subject of transmitting, I suggested we talk it over with Chuck.

A few weeks later Abbie came to our loft to view some tapes of himself speaking at a rally in New Haven in support of Bobby Seale. We had heavily edited his remarks, rearranged the order of what he'd said, but he didn't seem to mind. He just lay on the rug in the control room silently watching himself, a distracted look on his face. When the tape ended, he said nothing about the edit. He wanted to know about a guerrilla transmitter.

Chuck had enough technical knowledge to appreciate the outlines of the problem. He said what we would need was a modulator, a device that would change the video and audio signals from our portapaks and cameras into the type of electronic signal that could be received by a standard TV set. What's more, Chuck said, he knew where he could get one that might work. He made a call to his old employer and told us they had a used one they'd sell for $325. Videofreex didn't have near that much money in the bank, so Abbie wrote a check to the company while he lounged on the floor of the control room. He wanted us to buy the thing and set it up. My unattributed contribution to *Steal This Book* had bought us our very own modulator.

A few days later, we invited him back to the loft to see a demonstration of the modulator transmitting from Chuck's shop to the control room next door. Abbie looked annoyed. He might have settled for Manhattan, although he'd have preferred the five boroughs and the suburbs. What we were showing him was a toy with no relevance to his grand schemes. He demanded to know what more it would take. Chuck said he needed an amplifier of some sort, but he didn't know where to find one. The modulator we'd bought was commonly used in cable TV systems. A radio frequency amplifier, the other main component in a transmitter, fell outside Chuck's experience. He'd have to build one from scratch and he didn't know where to start. Abbie stalked out of the loft, cursing us.

∞∞∞∞∞∞∞

While Abbie's project excited us with its blend of drama and politics, it had no practical application in terms of our immediate and pressing needs. Our situation in the city was becoming more tenuous by the day, so we put the modulator on the shelf in Chuck's shop with our growing collection of broken, obsolete or otherwise useless equipment. We took it down again only once before we packed it for the trip to Lanesville.

The May Day Video Collective came together pretty much spontaneously. Its sole purpose was to videotape the 1971 May Day weekend anti-war demonstrations in Washington, D.C. That protest was a made-for-video event and drew students and independent video people from throughout the East to the capital. The leaders of the demonstration had planned peaceful, non-confrontational protests during the weekend. Then, on Monday, May 3, militants would attempt to close the government by blocking the major roads into the District of Columbia.

From the outset of the project, no one gave much thought as to how the May Day tapes would reach the public. That irked me. Without access to an audience the whole effort sounded hollow, as if the experience had to do with making tapes rather than getting the word out–an elaborate training exercise. I didn't need training. At a huge Washington rally protesting the U.S. invasion of Cambodia a year ealier, I had pointed my camera toward the line of police in front of the old Executive Office Building next to the White House and focused on a man in a suit wearing a plastic badge. He stood behind the police lines taping me with exactly the same type of portapak I was using to tape him. He would have an eager audience for his tape. I would have no such thing for mine.

For May Day, I had a plan. We would adapt Abbie's broadcast idea to the needs of the action, transmitting our tapes throughout the encampment of protesters. Chuck and I worked feverishly Saturday afternoon while demonstrators were listening to music and speeches on a nearby stage. We thought we could improve the performance of the modulator with a makeshift wire antenna, which we mounted atop a rented truck. Then we plugged a monitor into my old inverter and drove around the campsite to see where we could receive the signal. If it worked, we planned to put TVs at strategic locations around the encampment and broadcast helpful information, political diatribes, old tapes, whatever. Time after time, no matter what we did, we got nothing but snow on the screen. It didn't work any better than our experiment at the loft. I could sense why Abbie had grown so frustrated. The signal didn't have the strength to travel more than a few useless feet from the antenna.

The civil disobedience began in earnest just after dawn on Monday, a gray, chilly morning. Davidson had been arrested the day before during a police sweep near the Mall. No one bothered to confiscate his portapak, and he'd ended up with a prisoner's-eye-view of his D.C. jail holding cell gradually filling to standing room only with young, white demonstrators. The police let him out that night. The cells were filling up too fast. He chose not to risk rearrest Monday, as teargas began wafting over much of the central section of the capital. David was taping on his own on Monday when he recorded a demonstrator struck by a car whose driver clearly intended to run over any protester who got in his way. A cop wearing a helmet and high riding boots appeared from nowhere. Not stopping to aid the fallen pedestrian, he charged the camera, swinging his nightstick to disperse the crowd so more cars could pass through the intersection. You can hear the whack as his stick catches David on the back of his thigh. A few blocks away, a car full of men with short hair wearing shabby clothes and headbands vandalized trash cans not far from the White House. Surprised by a May Day camera crew, the men did not realize at first they were being taped as they spoke into handheld police radios. When they saw they'd been exposed as agent provocateurs, they jumped in their unmarked sedan and sped off.

On every side of us as we drove through Northwest Washington, cadres of D.C. police in riot gear and gas masks chased kids down streets and alleyways, tossing gas canisters after them, tackling and clubbing the unlucky ones. Those who escaped regrouped to heckle the police at a different location, tossing trash cans into the street or lobbing teargas back at the cops. To my surprise, I found myself tense but disengaged. For the most part, the scene looked like spring break for radicals. I searched unsuccessfully for signs of cohesion and hope. I could hear from the excitement in their voices that the protesters were having a good time the way post-adolescent mischief makers have always amused themselves–by taunting authority.

Something fundamental had gone haywire. People were dying on the other side of the world that day; and we had appointed ourselves to come to Washington to help end our government's role in the killing. But what I saw on the streets didn't look anything like an effort to do that.

Later that day, thousands of protesters gathered in front of the Justice Department, sat down and blocked the main entrance. From somewhere in the crowd a young man pulled out a flute and blew a Beatles tune. The protesters began to clap to the beat, a few at first, then everyone. The sound echoed off the stone building until the clapping overwhelmed the music. It was the lone effective act of mass civil disobedience during the whole demonstration. The cultural and political revolutions had converged, and here, too, the counter-culture had clearly taken the lead.

Exactly a year earlier a squad of Ohio National Guardsmen had taken careful aim across a campus lawn and fired on a group of unarmed student demonstrators at Kent State, killing four and injuring eight. None of us in Washington for May Day expected a repetition of that tragedy. The public outrage and the government investigations sparked by the Kent State killings had brought home to police forces around the country the self-interest of having a disciplined response to protests. And because the police had learned to behave better, we could afford to act worse.

Videofreex had come to May Day because our experiences in Chicago in '69 had cemented for the four of us who went there a sense of obligation to document the progress of the anti-war movement. By the time we decamped for Lanesville only a few weeks after the May Day demonstration, it felt to me as if the anti-war movement had splintered beyond recognition. Leaders and followers alike seemed exhausted and confused. I had no idea what contribution we could make with video toward ending the war or if we could make one at all. The mainstream media had begun its turnaround on the war, shifting with glacial reluctance from uncritical boosters to timid skeptics. The Chicago police riot at the 1968 Democratic Convention had startled many reporters. Kent State had rattled a middle class public that could too easily imagine its own children in pools of blood on a college driveway. Publication of the Pentagon Papers exposed the hypocrisy of historical rationalizations for the war. Insupportable body counts and the My Lai massacre case had played a role, too. Mylai had shocked almost everyone but the combat soldiers who served in Vietnam. The weight of the lies that sustained the war, the exposure of our government's inability to prevail–even to define what it meant to win–was building to a critical mass even as the movement was coming apart.

The anti-war movement as a phenomenon larger than uncoordinated bursts of campus unrest had played itself out by May Day 1971; the movement had ceded momentum to the corporate media, which embraced the anti-war cause in its fickle, myopic way. For us, that meant a few curious May Day Video Collective members showing up at the farm in the weeks and months following the Washington demonstration. Other than those contacts, we got no calls to come cover demonstrations. We drifted away from documenting the struggle we still believed in, not because of the distance between Lanesville and the centers of population where the movement still functioned or because we lacked commitment. We saw little happening on the domestic anti-war front worth putting on tape. The May Day edit, a compilation of sequences from everybody who had shot tapes for the project (the collectively edited final version was completed in the chaos of the loft at Prince Street as we packed for the move to Lanesville) had played here and there and then disappeared as old news. It could boast dramatic images of face-to-face confrontations with club-wielding police and fully armed marines leaping from helicopter doorways onto the Mall in front of the Washington Monument, images the networks hadn't captured or wouldn't show. But there was no audience for our inside view of the anti-war movement now that the networks had concluded the war was ugly and unwinnable and would portray it as such, despite the Pentagon's vain insistence that the old deceptions were necessary for the national defense. How odd, then, that the FBI should wait until this point, when we'd left the city for Lanesville and had so little involvement with the anti-war movement, to begin spying on us.

∞∞∞∞∞∞

It may have had something to do with a stately brick house on a quiet block of West 11th Street, between Fifth and Sixth avenues, in Manhattan. On March 28, 1970, an explosion in a basement bomb factory set up in the house by members of the Weather Underground demolished the building and killed three of the bomb makers, all of them young, white radicals. Two others escaped and went on the run, remaining fugitives for years. Other radicals had also gone underground and all of them were proving extremely difficult to find. FBI Director J. Edgar Hoover, as obsessed as ever by wild but often calculated overestimates of threats from subversives and saboteurs, instructed his agents to use the hunt for the bomb-factory fugitives as a rationale for expanding the government's already substantial efforts to spy on and harass those engaged in all manner of legal dissent. Although we had operated openly and actively for nearly two years in ways that might have attracted the attention of the Bureau, the dragnet Hoover cast was so large and far-reaching that it took his agents a while to identify us as a peril to democracy.

As Annie looked out the kitchen window one afternoon that first summer in Lanesville, she noticed a stranger. "There's a guy in the driveway taking pictures," she said. None of us knew the tall, thin man in the business suit snapping photos of our cars, and he hurried off before anyone could reach him. A few days later a green van, the same color as those used by the phone company but absent any markings, parked next to the telephone pole across Route 214 from the house. Davidson grabbed his still camera and went out on the front lawn to shoot some photos, then walked down the slope that led to the road and casually circled the truck. He came back to report that a man was sitting inside with headphones on. The truck left not long afterward and didn't return.

That summer was as beautiful a season as I can recall. It began in rain and mist, leaving the mountainsides surrounding us a deep, impenetrable green. We admired the scenery between runs to the city and to our Council-funded video projects in Rochester and around the state. We had successfully redirected our collective energies away from the political concerns so pressing in the city and applied them instead with an evangelical zeal to preaching the miracle of video to the citizens of upstate New York. With so much going on, we had no time for paranoia. The FBI remained, for the present, a villain of indeterminate presence.

∞∞∞∞∞∞∞

As summer ebbed some of us began to feel the absence of broadcast TV more acutely than before. We had attempted to erect a receiving antenna in a clearing a few hundred feet up the mountain behind the house, but that ended in a dismal failure. The signal couldn't make it down the mountain, so we temporarily abandoned that plan. We were left, for the time being, with our tapes, each other and our guests.

We had so many visitors we decided to restrict guests to weekends, though we didn't enforce that rule strictly. We wrote down who we expected so we'd have enough beds and clean sheets. He/she/they are your guests, so you clean up after him/her/them. We didn't allow anyone to crash on the couch downstairs. A bedroom or outside. Period.

Curtis met a stranger named Cyril Griffin. He had flame-colored hair and a beard to match. I don't recall now why he came to Lanesville, and I might never have known what brought him; we didn't always ask why people wanted to visit us in Lanesville, at least not at the beginning. What we knew about Cy was that after stints in advertising and design, he was a man between careers. He was just at the beginning of a long, passionate commitment to the cause of militant Native Americans, a path that led him into the midst of the federal government's violent siege at the Wounded Knee Reservation. But when he first came to the Farm, he seemed a hapless cipher, no job, no prospects, no video equipment, unreliable car, the type of person who frequently showed up hoping to feed off the energy of Videofreex without making any clear contribution of his own to the group. Cy's first wife had died a few years earlier, and he was caring for the two of his four children who were still minors. Some of us wondered whether he and his family were all about to become part of our household, which could have happened through the unwritten spousal membership rule. But Curtis ended the suspense late that summer when she announced she and Cy planned to marry and to live elsewhere. Their only request was that they have the ceremony at Maple Tree Farm and that I act as minister.

I need to explain this. When I lived in California, I heard about a man named Kirby Hensley, who had started the Universal Life Church in, as I recall, Modesto. An anarchist of a sort, Hensley believed ecclesiastical titles were poppycock. He chose to express this sentiment by incorporating his own church and then conferring the title of minister on anyone who joined. The way one joined was simply by writing him a letter requesting membership. I wrote–one of thousands who did–and received not long afterward a card with my name on it identifying me as a certified ULC minister. Hensley didn't care what his ministers preached or if they preached at all. He had an ecumenical outlook and a primary goal of devaluing religious hierarchies.

I used my minister's ID only once, when a cop stopped me on a cold night in Philadelphia for running a red light. I swore I hadn't. He looked me over: beard, battered leather jacket, hair getting long, driving an old car. "Wha'd'ya do for a living?" he growled.

I didn't have a job at the time so I blurted out, "I'm a minister."

He took a step back from the car. "OK, then. Just be careful," he said and walked away.

Curtis must have heard me tell that story, so what could I say? If they didn't want some turned-around-collar churchman, how could I refuse to marry them? Curtis had generously extended the hospitality of the loft she and David shared on the Lower East Side when I'd first come to New York City. I owed something to old Kirby Hensley, too, for getting me out of a ticket in Philadelphia. The wedding of Curtis and Cy that fall turned out to be the biggest party we ever had at Maple Tree Farm, and probably the largest in Lanesville for a long, long time. Art Varney, a friend of mine from Poughkeepsie, who had coordinated the Video Trips encampment at the Woodstock Festival, came to Lanesville several days before the wedding and set up a large tent he had liberated after the festival. A hundred people showed up for the prenuptial party the night before the ceremony. The house was full and most people camped in the clearing behind the houses next door. The party was a raucous affair that lasted all night. I made the mistake of drinking too much and then sampling some of Cy's peyote punch, which made me nearly crazy with a combination of hallucinations and hangover. I have no idea what shape anyone else was in, although I thought I had imbibed with restraint compared to those around me. It was my last experiment with psychedelics.

In the midst of all the preparations and partying, film maker Shirley Clarke, one of the guests, decided to direct her own video version of a Marx Brothers movie, using the party as a backdrop. She cast Skip as Harpo, Bart as Chico, and Allan Sholem, briefly a member of Videofreex in the city, in the Groucho role. Shirley had to be the center of attention wherever she went, never mind if it was somebody's wedding. I have always considered her an enormous talent, and over the years she taught me many valuable lessons about the uses of video. But she could also make herself an incredible nuisance.

At the party, and leading right up to the wedding, she set up shots everywhere, demanding her "actors" do impossible stunts on porch roofs and overcrowded hallways. Shirley's form of relaxation was hyperactivity, and she ran everyone around her ragged with manic requests. Her deep-throated directions, so oddly booming from such a slight person, heightened the overall pandemonium. Or maybe it was my own overall sense of pandemonium. I was having trouble maintaining my balance. I couldn't tell whether or not I was in her movie. I was in somebody's movie.

I still find the videotape of the wedding nearly impossible to watch. It reproduces pain in my head and stomach, my only memories of the event other than what I've glimpsed on the tape when someone else has played it. We gathered on the side lawn. Everyone sat down. I almost fell over. A stupid grin creased my face. I kept the ceremony brief and somehow managed to ask the right questions. I made several unsuccessful efforts to stand up. Uuuhhhh. I called for everyone who believed Curtis and Cy were married to make a sound, a clever way of shedding responsibility for the marriage by transferring it to the collective realm. The guests whooped or blew whistles, at which point I got to my feet unsteadily, followed by the newlyweds and all the guests and we danced around the side yard in a big circle while Bart banged on his pump organ.

The wedding was our first big video social at the farm, an apolitical event, as any wedding should be. It also served as a belated celebration of our arrival in the country. We had made a clean break with the city. We had our equipment in place. We had work to do as a group and our living arrangements, while not frictionless, gave us each enough space so we could function with a minimum of bickering. But the same problems that had dogged us in New York from the end of the Don West gig right through the May Day experiment, the questions of what to do with our tapes once we made them, and for whom we were making our tapes in the first place–those questions remained unanswered.

3
Achtung!

The popular press in the early 1970s bubbled in anticipation of the "video revolution," the imminent arrival of home entertainment systems, the marvel of having a world of programming options on tape or disk in your living room. Not just Hollywood movies, either. Savvy investors were reportedly lining up for the chance to get in on production and equipment schemes to feed this vast market. At the same time, anyone who investigated this new technology, no matter how cursorily, couldn't help but see its do-it-yourself underpinnings; for the first time people could make their own TV programming instead of having to consume what others produced. In a feature piece on the medium accompanied by a photo of a man with a portapak taping a naked blond woman stepping out of the shower, *Playboy* magazine gushed, "Video-tape recording may become the most popular hobby of them all."

Hey, wow! Here we were at the forefront of another revolution. All we had to do was define some recognizable territory between Hollywood and hobbyists to call our own. We got ourselves some slick stationery and wrote down the types of things about us and our services we knew those savvy investors would want to hear. Then we went hunting for money. The savvy investors probably had enough savvy not to return calls from a company called Videofreex. Or maybe they'd taken a closer look at the "industry" and discovered that other than educational films on tape, it didn't exist.

We found a few high-profile jobs after our break with CBS, including the projection of video images for the New York Pop Festival at Randall's Island Stadium in New York City. We ended up sharing the stage with performers like Van Morrison. But to get paid for that shoot, we had to stage a sit-in at the posh midtown Manhattan offices of the concert producer, who grudgingly gave us a few hundred bucks to get our scruffy selves out of his glass-enclosed waiting room. A few hundred bucks hardly paid for the tape we'd used.

We recorded hours of live performances by big groups and unknowns, all mixed through our rudimentary Sony special effects board. Our hand-held, whirligig, multi-camera style consisted of odd angles and extreme close-ups, with switches, wipes and dissolves between shots paced to the beat of the music. It's now the accepted production standard for the immense music video business, not to mention TV commercials. But MTV, VH1 and all their spin-off music video channels lay years in the future. The people who controlled the gateway to broadcast television, the only mass distribution medium that existed for videotape, scoffed at what we did. Sloppy camerawork, they said. And black and white–nobody watches that anymore. Too grainy. Too jangly.

We never got to the point of having to admit we held no broadcast rights to the music we taped. So few people had VTRs, there wasn't even a bootleg market. Not until the beginning of the '70s did the manufacturers of VTRs using half-inch wide tape come to an agreement on a standard so that tapes made on one company's machine would play on the VTR of another manufacturer. And that agreement didn't affect more sophisticated VTRs using wider tape. The word "cassette" had not yet entered the video lexicon.

∞∞∞∞∞

In those first few years, our most popular tape turned out to be one with no music and little apparent commercial value: our interview with the late Fred Hampton. During the early 1970s we could sense from the morbid glee of mainstream media and from our underground sources the Black Panther Party was losing its struggle to become the militant voice of the nation's major black ghettos. The Panthers were never a large group to begin with, and the personal flaws of their leaders have now been well documented. But they didn't have a chance to test the strength of their 10-point program for community self-rule as a remedy to the oppression of poor African-American neighborhoods. In addition to the overt hostility of fearful and racist local authorities, the party's principal enemy was the Federal Bureau of Investigation. No one outside the warrens of government knew at that time the extent of the FBI's activities, but the Bureau had targeted the Panthers for destruction, and the Bureau's disinformation campaign had, by this time, successfully exploited the party's internal weaknesses, helping split the Panther organization into hostile factions.

About a year after Hampton's murder, Big Man showed up at our SoHo loft, his black leather jacket strained around his waist. Two men he did not identify accompanied him. They watched the Hampton tapes in silence. I want you to play these tapes for the people, he said, when the last tape stopped. He waved off our explanations about broadcast obstacles.

Big Man headed the party's New York chapter–or one them. Egged on by the FBI and abetted by a reflexive willingness to demonize the Panthers, the mainstream press trumpeted accounts of violence between Panther factions without bothering to investigate the underlying causes. Most of what we knew about the split came from the press, and we had no way to assess what it meant for us to work publicly with one element of the party, whether that amounted to taking sides, and if so, whether that might prove dangerous to us. But we didn't see how we could refuse Big Man's simple request that we bring a monitor and VTR to a community center in Harlem and, later, to another enthusiastic audience at a drug rehab project on an island in the East River. We were met by a Panther escort each time, which didn't make me feel any less uneasy about being there.

We'd recorded Hampton with a wide angle lens at the home of a wealthy Chicago woman named Lucy Montgomery. She owned a Prairie School house furnished with modern art. Hampton arrived late with a small entourage and paid no attention to the lavish surroundings. He looked tired but strong. He was chairman of the Illinois chapter of the party, and though he was just my age, he seemed much older to me. If our crawling around to frame him from all different angles bothered him, he didn't let on. He had a message to impart and ignored the distraction. On the tape, his deep voice rolls out across his hands, almost as if he's using his practiced gestures, so often in the foreground of our shot, to impel his ideas onto the screen.

In a long-winded way, I ask what would happen if he is killed. No hesitation. He says the party is stronger than any individual and the work will continue.

Hampton yields occasionally to a member of his entourage, a thin man named Omar. Unlike the other Panthers in their austere black jackets and berets, Omar wore a funky, broad-brimmed hat. He delivers a whiny, high-pitched rap, mostly jive, and out of courtesy we taped far more of him than we wanted. Big Man said Omar was a police informer. He said it as a matter of fact, a clear statement that we were not to play any segments of Omar except where it couldn't be avoided. No one wanted to hear Omar, anyway. But people would sit forward quietly and concentrate as Hampton spoke.

Not long after the New York showings of the Hampton tape, a French woman, white and impatient, dressed better than a hippie and identifying herself only as Lily, appeared at the loft. She had a taped statement from Panther leader Eldridge Cleaver, the party's original minister of information, and next to the poster image of Huey P. Newton holding a rifle and spear, the face of the Black Panthers best known to white Americans. In prison, Cleaver had written a book, *Soul On Ice*. It had sold well and received critical praise. The attention paid by the mass media to the very fact he had written a book, perhaps more than his perceptive political analysis of contemporary American society (or the homophobia or quirky sexism expressed in *Soul On Ice*), imparted celebrity status to Cleaver. Lily had brought the tape from Algeria, where Cleaver was living in exile. She wanted it broadcast. We told her what CBS and the other networks had told us, that our half-inch tapes could never be broadcast because they would cause the picture on viewers' sets to break up. Lily insisted we try.

No one at CBS knew Bart; he'd had no association with Don West. So he agreed to escort her to the Broadcast Center on West 57th Street that afternoon. Once they saw the tape, the network's news producers wanted it, and within minutes engineers found a way to play it on the air. That night, a long excerpt of Cleaver led the CBS Evening News with Walter Cronkite. It was the first network broadcast of half-inch video and there were no technical repercussions reported.

Years later, a sequence from the Hampton tapes also found its way onto national TV as part of the PBS documentary series on the civil rights movement, *Eyes on the Prize*. Those same tapes were used in the civil case Hampton's family brought against the Chicago authorities for his murder. They were played to the court to help establish that Hampton fit the profile of the "black messiah," an imaginary figure J. Edgar Hoover invented and then came to fear so desperately he sought to eliminate all possible candidates. The case dragged on for thirteen years, with various setbacks along the way. But the families of Fred Hampton and Mark Clark, who was murdered by police in the same raid, finally prevailed in a case that established the authorities had violated Hampton's civil rights.

Our direct contact with the New York Panthers ended after the showings of the Hampton tapes. Questions of survival overwhelmed them, and we had no help to offer. Our association with the party had consisted of the Hampton tapes and our willingness to play them in public; yet the fact that we'd had any involvement with the Panthers enhanced a growing mystique around the almost exclusively white, middle class video movement that Videofreex was the radical video group, the one out there on the edge of the struggle for peace, freedom and justice. Compared to many of the other video groups in New York at the time, perhaps the perception was correct. We certainly did nothing to discourage the notion. Legends can prove useful in establishing an identity. Legends are also a poor substitute for history. Our involvement with the Panthers' fringe of the civil rights movement was peripheral, consisting mostly of transactions at what for us was a safe distance. After we moved to Lanesville, we read in the *Post* that Big Man had been murdered.

Lily disappeared as suddenly as she arrived. I never saw her again. But copies of the Cleaver tape she left us found their way into some of the later showings we did for the Panthers. Together, the tapes took on an identity of their own. They were called the Voodoo tapes.

From a technical standpoint, Lily's visit had been instructive. We always suspected the networks were lying to us about the technical barriers associated with broadcasting half-inch tape. They had erected a wall of engineering gibberish to pre-empt any demands that they air our programs. The Cleaver Voodoo tape confirmed the deception. It was possible to broadcast half-inch tapes. We had unmasked the pretense. We had not, however, broken down the walls of resistance.

∞∞∞∞∞∞

Most mainstream reporters and editors in the early '70s chose to stress video as a new artform and downplay its political implications and its potential impact on the extraordinarily profitable broadcast business. The origins of this attitude stemmed in part from a long *Newsweek* story on video in February 1970, the first big piece on the medium in a national magazine. It was written by Douglas Davis, who'd found Don West the fall prior to the story and wangled from him the chance to shoot a tape. He'd chosen to record Carol sitting on her bed in her Manhattan apartment, talking to him as he stood behind the camera. Slight problem, though. No sound.

Don thought he should play Davis's screwed-up tape as the lead-in to our show for his CBS superiors. He knew it was crap, but how could he afford to thwart the ambitions of an influential critic at a powerful magazine? Fresh from his initial video triumph, Davis announced he had now become a video artist as well as a critic. Given the credentials of the rest of us, who could argue? And where *Newsweek* led, others soon followed.

The Videofreex, and those of our contemporaries who held strong political views, about the media earned only passing mention in most of these stories. Davis, for instance, used most of his story to describe projects funded by the Ford and Rockefeller foundations at public TV stations WGBH in Boston and KQED in San Francisco. Our more radical message that corporate media monopolies suppressed the voices of common people didn't sit well with the very folks whose jobs we called for eliminating. Anyhow, we amounted to small potatoes by mass media standards. Davis finished his piece with a few sentences on how CBS had canned Don's show, concluding, "The real future of TV as an independent art form rests, therefore, mainly in the hands of its fringe practitioners ... ." But that didn't mean the mainstream press would devote much ink to us "fringe practitioners" and our vision of an egalitarian media future in which everyone would have access to the airwaves. It was a justifiable bias considering we hadn't accomplished enough to give us political weight or social standing. We didn't hang out with stars, either, or factor in any juicy scandals.

A writer for *The New Yorker* magazine had stopped by for a show at the loft. He sighed in his piece that our style brought to mind an updated version of his "Paris-in-the-twenties" fantasy. He offered a brief, factual description of the tapes we played that night, including Hampton, and then quoted me describing Videofreex: "We are a bunch of struggling crazies." Not exactly revolutionary rhetoric.

A decade or so later, the magazine would return to the subject of video for a major story proclaiming Korean-born artist Nam June Paik the "George Washington of video," for his early experimentation in a medium that had by then become an established artform. Nam June is a joyful, endlessly inventive and engagingly loony man, a fixture in the video world. Among his best known pieces is TV Bra, in which the late cellist Charlotte Moorman, topless, played her cello while wearing two small monitors on her breasts, generating squiggles on the screens with her music. His recorded pieces rely on playful mixtures of color and sound so non-threatening to the established powers that he was allowed to broadcast his *Electronic Opera No. 1* on the Boston public TV station in 1969. It was probably the first time any station admitted to airing video art.

Unlike our confrontational attitude toward the TV industry, neither Nam June's work nor his personal style ever directly challenged broadcasters. As exacting as he was in his work, he always displayed unflagging cordiality and generosity to those around him. Once, during a video and film exhibition in Belgium, he loudly proclaimed me a "genius!" for discovering that the circuit breaker on his monitor had popped off. I reset it and over and over he proclaimed my brilliance in his high, thin voice. But despite his lofty opinion of me and Videofreex, we could never persuade him to visit us in Lanesville. He worried that once he left the big city he'd be mistaken for Japanese and killed by American veterans of World War II living in the hinterlands. It's the one political statement I associate with Nam June, who certainly never discussed his art in political terms. Nam June's work was marketable because his message was considered safe.

Our radical image made us unsafe. No one in the straight press would willingly give a group like Videofreex a forum for our political views. That proscription, however, did not apply to the topic that most intrigued the straight world: our lifestyle. After we moved to Lanesville, a small wave of reporters followed us, professing to want to look more deeply into what was happening and who we were. They came for a peek into the counter culture on behalf of the rest of society, voyeurs on assignment.

We must have given some of them a better line than others. *Horizon*, a pricey, hardcover magazine published by American Heritage, reported, "The shaggy technicians ... who call themselves the Videofreex, are dedicated to learning how to integrate technology and the human species, to form a community in which man and the machine can live in perfect, intimate harmony."

Ayyyy! We had no illusions about living in "perfect, intimate harmony" with each other, never mind our equipment. But we gradually came to realize that unlike people such as Paik, who fit neatly into the traditional mold of the individual artist, our video collective–we never used the word "commune" to describe ourselves–represented a phenomenon at once titillating and ominous to those on the outside. We didn't fit preconceived notions of how individual artists worked and neither were we employees of the television industry. By default, then, we must be a cult.

I find it hard to believe the way we lived lent much credence to the cult theory, but straight reporters couldn't help but speculate. The principle members of the high profile groups in New York City, Raindance, People's Video Theater and Global Village, all lived separately at first, although PVT members shared a loft for a while. Their private lives had a conventional cast that never came up in the reporting on their video activities. And small wonder. In the city, where people live so tightly packed and at such a high cost, living arrangements seldom stand out unless someone intentionally makes an issue of them. But in the country, where, theoretically, each person can have his or her own space, the way we lived took on an importance in the minds of writers and editors that frequently outstripped the work we did.

Over the years we lived at Maple Tree Farm, we never hid any of our activities, with the occasional exception of the people who smoked grass. Strangers could and did drop in unexpectedly, although we quickly got to a point of discouraging that practice so we would find time to work. We asked visitors to arrive no earlier than Friday afternoon and leave no later than Monday morning, unless they had come, by prior arrangement, to work in the studio or for other business. This rule did not apply to family members or local people, and neither of those groups abused the privileges of the house, except for Sam, our landlord, whenever he could slip out of his house without Miriam noticing.

We had our share of parties, though nothing other than the wedding ever bordered on ritualistic behavior. Our biggest celebration other than the wedding was a nominal "luau" on the banks of the Stony Clove Creek, where Davidson roasted a deer hit by a car. We invited everyone we knew in Woodstock, and plenty of people from the city. A few semi-indigenous bikers showed up, too, and pronounced themselves satisfied with the level of noise and intoxication. It became raucous by any standard, but hardly cultlike, unless you read some significance into the drunken drumming and the howling of Grateful Dead songs, which lasted well into the night.

A work ethic prevailed at the farm, no matter how stoned some folks got while working. To relax, Videofreex and guests sat around the kitchen table after dinner, sharing the last of the hippie beer, sometimes a joint, and talking long into the night about video projects, or moving to the viewing room to watch tapes. We showed our visitors to one of the many guest rooms and otherwise left them to snoop for themselves. No one proselytized except on behalf of the virtues of technology, and reporters got as many versions of the party line as there were Videofreex.

But if you insist on looking for a cult, you need first to identify a leader; and sometimes David willfully failed to disabuse reporters of his fantasy that he filled that role. The *Newsweek* story had confirmed our worse fears in this regard, when Davis wrote Videofreex was "led by artist-writer David Cort ... whose video journalism has been shown in galleries and museums, as well as on his own closed-circuit loft system." I can still hear our cries of horror when we read that, as if David and Doug Davis were in cahoots to paint the rest of us into the background. But it wasn't entirely David's fault. For one thing, the mass media needed to identify a leader. Reporters like Davis seemed incapable of understanding that late-twentieth-century Americans could work together in ways that didn't require a hierarchy. We might say Videofreex didn't operate on the military/corporate model, but how could that be? How could people–especially artists–voluntarily cooperate? Gotta be a leader.

And you had to take into account David's irrepressible self, too, with his penchant for theatrical bellowing, his wheezy laugh, more a tic than an expression of humor, his Whitman beard and Afro hair, and his parrot, Oberon, sitting on his shoulder nibbling his ear. David presented more colorful character to the press than the rest of us, and his willingness to accept the mantle of leadership rather than discourage the perception that he spoke for the group left us seething whenever those misconceptions appeared in print. It happened often enough that we became wary of leaving him alone with reporters, so much so that at times it may have seemed as if we exercised a form of group censorship. But it wasn't censorship so much as a desire on the part of each of us to be recognized and heard.

It wasn't only the print media that made us wary. Valeska and her crew from German TV showed up in Lanesville the day after the big snowstorm our first Thanksgiving at the farm. She carried herself like a dancer and wore her brown hair cut tight to her head. She asked questions as if giving orders. "I think you can answer what video means to you, yes? Just go about what you do, yes? We film you as you are."

Constanza, her diminutive soundwoman, constantly shook her head either from technical difficulties or because she simply couldn't believe what she was hearing through her headphones. I never knew which. Albert, a towheaded athletic man, operated the camera. He clearly enjoyed peering over our shoulders through his lens.

They had come to film us, and at that time the medium of film enjoyed a prestige denied video. As a practical matter, film images possessed a clarity and crispness unattainable with early half-inch video; film had color and we could shoot only in black and white; the sound on films often came out more distinctly, though usually because film professionals took sound recording seriously and used better microphones rather than because of any technical inferiority of videotape sound. And most important, broadcast stations and theaters would buy and play films. Not so with videotapes.

At the height of the student movement against the Vietnam war, film was the alternative visual medium. Videofreex could offer little more than our services as archivists and ego strokers. Films played on every college campus, whereas VTRs, when they existed, were locked in a closet of the college athletic department. Sixteen millimeter film had become an international standard, while almost no one in Europe could play videotapes made in the U.S., and vice versa. And film tradition had an established creative hierarchy, with the director at the top. We liked video precisely because we could do it as an undisciplined team effort predicated on anarchy.

When David and I paid a courtesy call in 1969 to the radical film makers' cooperative called Newsreel, whose New York chapter was located in an old firehouse on the Lower East Side, we received a decidedly cool reception. The Newsreel folks managed to achieve something years away for video: an alternative distribution network which, though shaky, actually did manage to get radical films to receptive college audiences. And even if video had been a viable option, most people would prefer to gather around a film screen rather than a crummy, blurry, 17-inch black and white monitor with a tinny, five-inch-wide speaker, never mind that most students, then as now, spend a lot more time in front of the tube than in a movie theater.

We willingly conceded these points, and still we had no respect for film as a culturally superior medium, nor for the snooty attitudes of film people. We didn't bother to hide our bias any more than film makers, by and large, disguised their disdain for our medium. As a matter of principle, we saw film as elitist. Film was extremely expensive to process. A canister of sixteen millimeter film only lasted eight minutes as compared to the thirty minutes we could record on our portapaks, re-recording over any tape we didn't choose to keep. And there was no way film could be shown immediately. Film was the past, video the future and we let that German film crew know it right away.

The first sign of trouble came when we pelted Albert, the cameraman, with snowballs while he was shooting. He didn't complain, but Valeska was not amused. "No, please! You are not to acknowledge our presence here, yes?" A snowball hit her square in the face. She ordered the crew inside. Constanza shook her head.

Because this was Thanksgiving weekend and we had no pressing deadlines, we could devote all our time to the playful torment of our visitors. It was one of the few instances during all the years at Maple Tree Farm we experienced such unity of purpose. We agreed later it had something to do with Valeska's insistence on using the German word "Achtung!" to begin each scene.

Film makers and video people have radically different styles working. We would turn on our cameras and enter a scene, sometimes observers, sometimes participants, usually some combination of the two, but almost always with the idea of allowing events to unfold at their own pace. Film, however, ruled as it is by time limits and cost, imposes a formality consisting of structured scenes of roughly predetermined length. "Achtung!"

All weekend, whenever they filmed us, one of us was taping them filming. Sometimes, we taped us taping them filming us. German TV had an appetite for American culture, especially its fringes, but this was getting out of hand. So Valeska decided on a frontal assault. She herded all of us into the viewing room and ordered us to sit in what she determined was a relaxed and natural fashion: two rows, one row sitting, one kneeling. "Now you will all say a sentence about what you think is most important, yes? One at a time please, beginning with Nancy ... Achtung!"

The previous summer, we'd eaten together most evenings, and those collective meals had drained our enthusiasm for communal togetherness. Since then, we seldom gathered in the same room at the same time. Valeska either hadn't picked up on this or didn't care. The room was silent except for the whisper of the film grinding uselessly forward. Then a burst of uncontrollable laughter. Valeska fumed in German. "Cut!"

She pointed at me, the ringleader in this case. "Achtung!"

"We've had bedbugs recently. I hope you didn't have any ... " Constanza lifted her headset and inspected the earpieces. Valeska stopped mid-sentence, resigned to our unwillingness to collude in her efforts to force spontaneity. She had come to the wrong place for a scripted documentary. We had embraced our own adaptation of Heisenberg's uncertainty principle: When the media observe an event, the act of observing invariably affects what's happening. To pretend otherwise perpetrates an illusion, if that's the intent, or a lie, if it's not. From the experience of shooting countless hours of tape, we had learned the futility of trying to mask the camera's involvement in the scene. This was not a group policy. It was a shared perception that grew from our work. To her credit, Valeska gave up.

Somewhere in the archives of a German TV network is film of the Videofreex at home. We never saw it. Somewhere in the Videofreex archive is a tape of Valeska and her crew as they packed their bags, leaving earlier than expected despite the snow. We may have watched our tapes of them that weekend. Hard to believe anyone has viewed them since.

4
Do
They
Mix It Up?

What were we, then, if not a cult or a commune? We dodged that question with our vague, reflexive answer: A collective. We're a video collective.

A what? Did we mean artists? No, too pompous for me. But group wisdom prevailed. Don't be silly. Everything we do qualifies as art. Arts funding pays the rent. A tape of us eating dinner: art. Chuck's workbench: art. Polaroids of Davidson and a slender young woman posing formally on his brass bed, a supercilious look on his face, she, demurely in his arms, a photochemical Vermeer, except they're both nude: art. Unfinished anything: art.

And perhaps technoids, too, to outsiders. But most Videofreex didn't know one end of a soldering iron from the other. What, then? Ideologues? We had no shared religious or philosophical belief around which to organize our lives, nor did we owe allegiance to a single benefactor, excepting our uneasy obeisance to successive bureaucrats at the State Council on the Arts. We didn't fit neatly into any historical tradition.

Not much in the Greene County Catskills harked back to collective roots, either. In the mid-1840s, Hudson River School painter Asher Durand proclaimed his artists' colony in Palenville at the foot of the Great Wall of Manitou, the first such colony in the United States. Durand's art is still admired; his colony is a dusty footnote. More than a half century later, Hervey White founded the Maverick after he'd broken with Ralph Whitehead's Byrdcliffe arts and crafts colony in Woodstock. Byrdcliffe rested on Whitehead's elitist vision, and he alone bankrolled it. The Maverick community, as historian Alf Evers describes it, was best known for its annual, freewheeling operatic revelries known as the Maverick festivals. But the Maverick had withered more than a generation before we arrived.

The advent of Videofreex as a group sharing common living facilities was preceded about a year earlier by the Rainbow Farm in Woodland Valley outside Phoenicia–about a dozen adults in and around an old clapboard farmhouse high up a mountainside. Some Rainbow farmers ran a food shop in Woodstock. Most of the members of the group were building their own houses on land that belonged not to them but to an absentee heiress.

46

For a while, David attended Rainbow Farm sweat lodges and tepee gatherings. Their more laid back approach to life and their involvement with the land rather than technology provided him a respite from the comparatively high-pressure creative atmosphere in Lanesville.

A group called the True Light Beavers lived one mountain closer to Woodstock than Videofreex. The core members were three brothers, Alan, Martin and Tobe Carey and their wives, plus several friends. Tobe was a film maker seduced by video, Alan a photographer, and Martin, a painter. They lived as an extended family, and as Tobe grew more involved in video, he became a frequent visitor at the farm and eventually a collaborator on a number of projects.

Brothers in one group, homebuilders and entrepreneurs in another. Completely understandable. But video freaks in the country was anything but a natural fit. Try the label techno-kibbutz. But, no: We shared no common faith or cultural ideal. Our beliefs ranged from skeptical to agnostic. Except David. He pursued contacts with Jewish organizations as he explored the first signs of his own awakening of religious commitment. Within the group, though, he didn't proselytize. He knew better than to try.

Our one shared principle was the credo of the workshop–an evolving belief that placing video cameras and VTRs in the hands of ordinary people would make the world a better, more just and beautiful place. Any time a member of Videofreex, usually though not necessarily accompanied by a portapak, encountered another human being, that encounter was recorded as a workshop and dutifully reported to our state benefactors. Driving back roads to Rochester for a big workshop, a handful of Freex see the Purple Lancers Drum and Bugle Corps practicing in a rural A&P supermarket parking lot. They stop the van, tape a few minutes, play it back, show everybody how easy it is, hop back in the van and drive off. A hit-and-run workshop. Hey, wouldja lookit how easy it is to make TV! How come the channels we get don't do this? A headcount of eighty for the arts Council, an audience invisible to a television industry that requires at least three, maybe four or five zeroes to the right of our numbers before they matter to advertisers. Grayson had explained it: It's not television, it's video. Video freaks refined the explanation: ... and it's subversive.

Sometimes workshops, sometimes art. Often the distinction disappeared. Davidson pursued a sculptural aesthetic. In one of his early pieces he pointed three cameras in slightly different directions toward the rim of Hunter Mountain, recording the images from each camera simultaneously on separate VTRs. When he synchronized the playback on adjacent monitors, he created the videotape equivalent of a panoramic photograph, in which, occasionally, a bird would fly or a cloud drift from one screen to the next. This type of image, now produced digitally, has become a common, eye-catching backdrop for TV magazine shows, a throw-away effect that buries aesthetic opportunities within a ponderous grid of monitors. David explored skewed perceptions of self and body through special effects. You could find your every movement mysteriously integrating some part of your body into a great painting, giving you a workshop in "virtual unreality." Skip, Bart, Nancy and Carol concentrated on documentary footage, with Skip, after he left the group, winning three Emmy awards. Annie steadfastly stuck to her watercolors and ink brushes, Chuck to his oscilloscope. My efforts ran more toward scripted productions and live performance. And for all these divergent interests, we still managed to live and do our work and our workshops together at Maple Tree Farm and on the road as long as we resisted the pressure to define ourselves too narrowly.

∞∞∞∞∞∞

Sam, however, had us pigeonholed within a few weeks. David was the "big philosophe," sometimes "nuts" as in, "Oh boy, dis guy is a big nuts," and sometimes "de boss," until Carol disabused Sam of the notion that David carried any special authority, at which point Sam conferred the title on her.

Sam and Miriam had bought Maple Tree Farm in the 1950s. Miriam ran it as a boarding house during the summers, with their only child, Leo, helping out when he was young. Sam stayed in Brooklyn at his union carpenter's job, and came up on weekends. They closed the place each fall after Yom Kippur and returned to their apartment in the city. When Sam retired in the mid-'60s, he and Miriam moved into the main house, with its 17 bedrooms and two kitchens. But the summer tourist business had dwindled over the decade following their purchase of the farm, their regular customers too old to travel or else well-off enough to take more exotic vacations. Leo had his own family now and couldn't help out the way he used to. So to make ends meet, Sam and Miriam rented the place for the winter to a ski club from Long Island and moved to a cottage just across the driveway in the back.

Skiing–these skiers–got us the place. Commercial skiing in New York began in Phoenicia in the mid-1930s. In 1960, the Slutzsky brothers of Hunter, Israel and Orville, opened the Hunter Mountain Ski Bowl at the north end of the mountain, and Lanesville people flocked to the slope, not to ski but to work. By the late '60s a few enterprising contractors began to build ski houses in the hamlet, so there were construction jobs for a while, too. Once complete, many of the new homes languished on the market, too far from the raucous nightlife springing up along the main streets of Hunter and Tannersville.

Sam and Miriam liked the money the ski club paid them to rent Maple Tree Farm half the year. But the debauchery they witnessed in their own home on ski weekends soured them on the prospect of another year with the club. Sam would curse the club members whenever we talked about them. He wouldn't tell me exactly what he and Miriam had seen except to shake his head and lament the drinking, the "daaaancing" and the "goils mit no clothes on."

Sam was a short, stocky, voluble man with a round face. He looked like a cross between former Soviet Premier Nikita Khrushev and Santa Claus without a beard. He grinned easily and sputtered saliva when he argued, which he did often. He was thrilled at the prospect that a group of "radicals," a term of honor in his lexicon, would displace the apolitical, lascivious skiers.

Miriam, shorter, rounder and more even-tempered than Sam, had a skeptical nature. She overcame her reservations about us by keeping in mind our promise to pay rent year round. At the end of June 1971, as the ten of us moved in to Maple Tree Farm, Sam and Miriam moved to a small house they owned just up Route 214.

Miriam warned Sam over and over, "Sam, don't bother the people," but he wouldn't listen. Using whatever pretext he could find, usually maintenance, he'd stride up the driveway to the kitchen door. "I can come in?" he'd ask. The door was open to everyone else. There was little we could say.

Sam absolutely adored Carol. She never hid her political views in conservative Lanesville. He, on the other hand, had spent years playing down his left wing sympathies when talking to his neighbors. He had lived through the violent organizing of the labor movement in the 1930s and the Red Scare persecution of the '50s. Those memories and an awareness of the anti-Semitism just under the surface of Catskills life, led him to avoid political discussions unless he knew his audience or the topic turned to Nixon, a man for whom Sam could not disguise his distaste.

He called Carol "mine Kaaar-erl" and never mastered her last name, Vontobel. He settled on "Vonlorber," which he used when he wanted to deliver a stern message on some household matter: "See here, Mrs. Vonlorber, the dog got to be on d' chain." This technique never worked in all the hundreds of times he tried it.

"Sam, that's bullshit," she'd say.

He'd turn red and revert to, "Aaaahhh, mine Kaaar-erl." He accorded her an extra measure of respect because she did the books and signed his rent checks.

Despite his deference to Carol, Sam had a difficult time accepting the equal contribution women made to the group, and frequently referred to Videofreex women as "nice, good goils." Eventually he lumped Nancy and Ann into the classification he reserved for Bart, Skip and Davidson: "artiste." He called Chuck by that nickname, or "Meester Kennedy," only occasionally bestowing on him the honorific, "woiker," or "woiking man," as in, "Meester Kennedy, you're a woiking man, not a boss."

He loved Skip and called him "Skeepy."

"My name isn't 'Skippy,' Sam, it's Skip."

"Awright, awright. I call you Meester Skeepy!"

For me, "mine friend, Meester Paaaaary," he reserved the highest praise, although I didn't see it that way at the time. "You are an in-tel-lec-tu-al," he would say, articulating each syllable. I would have preferred "artiste," but in a strictly relative sense, he probably had it right.

∞∞∞∞∞∞∞∞

Videofreex had ceased to exist as a company by the time we moved to Maple Tree Farm. We still called ourselves–and were known in ever widening video circles–as the

Videofreex, shortened frequently to "the Freex." But Media Bus, a non-profit corporation, was the key to our new strategy of trying to survive on government grants. We earned $25 a week above room and board until the end of each grant cycle, when the money ran out. All the Videofreex became members of the Media Bus board of directors, except Chuck, whose run-ins with authority made him wary of signing anything official.

We agreed Skip, not David, would become president. Carol, who had once taught at the same Harlem elementary school as Skip, had brought him into the *Subject to Change* office. After the failure of the CBS show, he'd bought his own portapak and claimed his place in the group. His high office at Media Bus–a strictly ceremonial post–reflected, in part, his having helped the accountant with our application to the IRS for tax exempt status. He was the best qualified because he'd completed a year of graduate study in business at Cornell (during which he'd also toured with a doo-wop group, lip-synching popular top 40 radio hits to screaming teenagers in the hinterlands). Except for Skip and Carol, the rest of us couldn't bother ourselves with paperwork

∞∞∞∞∞∞

We knew conventional broadcast outlets were closed to us and had learned how pointless it was to hustle commercial jobs that didn't exist. So it wasn't much of a stretch to conclude collectively that we should try instead to support ourselves as artists, temperamentally ill-equipped for making money. But in order to jump on the gravy train of arts grants Rockefeller had set in motion, we needed federal non-profit status. That prospect worried Noah Kimmerling, our accountant. Noah did the books for a number of radical groups and individuals, including Abbie Hoffman, and he thought the IRS might reject or delay granting Media Bus tax-exempt status because of our politics. That type of harassment was illegal, but there was little we could have done had the IRS targeted us. And yet, if we didn't get that non-profit designation, we couldn't stay in Lanesville or remain a group.

When we met with Noah at Maple Tree Farm late that summer of 1971, he proudly announced the government had approved our application. "You know the only thing the IRS wanted to know?" he said, nodding his head. "They asked me: 'Do they mix it up up there?'"

Everybody else laughed, but I didn't get it. Did we "mix it up?" Money? Fights? Now they laughed at me.

The Internal Revenue Service had no questions about our qualifications as a nonprofit organization, but the revenue agents had an intense curiosity about whether we swapped partners or indulged in group sex. They'd heard of hippie communes, and we came about as close as they'd ever get to one. Did they have a special form for "Sexual Proclivities of Applicants"? And exactly how would it have affected their ruling on our application if we'd satisfied their fantasies?

Noah wisely let the relevance of the agents' question go unchallenged, pleading ignorance. It worked. We'd never made a profit, had no plans to make one, and now it was official. For all we cared, the IRS agents could let their imaginations run wild about our living arrangements.

∞∞∞∞∞∞

We had picked out our individual living quarters on the second or third floors before we moved into Maple Tree Farm. The boarding house layout made sense for us. The couples would each get two adjoining rooms. We would share the kitchen, the largest room in the house, and work out a cooking schedule. The second kitchen, separated from the main one by a narrow pantry, stayed dark and chilly even in summer, and it became the equipment storage room. Another room off the kitchen became the office and later, Chuck's shop. On the remainder of the ground floor, the old dining rooms–two adjacent parlors separated by sliding French doors–became the editing/control room and the studio. The other small parlor on the opposite side of the stairway became the viewing room. That left the main parlor for what we called the music room.

Chuck hung our huge, homemade speaker cabinets from the ceiling. Our combined record collections filled one wall. Bart bought an old pump organ at an auction, a dark, ornate wooden cabinet with a row of stops and an ivory keyboard. It remained the centerpiece of the room long after its bellows gave out. A sun porch opened off the end of the music room opposite the studio, an unheated space with tall windows on three sides. All that first summer, we ate dinner together around a big table on the sun porch.

We had grown to ten people by then because Davidson had a companion, a poet named Frances Wyatt, whom he had introduced to us as we moved in. Her arrival went almost unnoticed in the hubbub of the transition to the country, and no one would have challenged the choice of a mate by an established member of the group. But the arrival of Frances marked something of a milestone. In the year-and-a-half since the end of the CBS gig, we'd only added three other people: Skip, Bart and Ann; and two of them, like Frances, had arrived as mates of people already in the group.

Skip had made a place for himself. He didn't need any sponsorship. After *Subject to Change*, he'd moved into one of the many bedrooms in Nancy and Carol's apartment on the Upper West Side of Manhattan. Like Davidson, his acceptance into the group was conditioned on his willingness to share his portapak without restrictions. It didn't matter anymore that he'd worked for Don; so had Carol and Nancy. When we got to Lanesville, he chose a room on the third floor, where most of the single people lived. Those rooms didn't share doors.

∞∞∞∞∞∞

Bart was driving a cab in early 1970 in Manhattan when he picked up Ira Schneider, a film maker, video artist and friend of ours. He asked Ira what he did for a living, and Ira said he was "into video."

"Me too," said Bart. "I'm a member of Videofreex." Ira knew us all and had never seen Bart before. Neither had we, except for Nancy, and she hadn't had any contact with him since the night a few years earlier when she and Bart had a date but she stood him up and went to Maryland, where she married someone else. Ira told us the story of Bart claiming to be a member of the group just before Bart called to say he was coming by the loft to offer his services. I didn't want anything to do with him or anybody else who had the audacity to masquerade as a member of our group, but I agreed to be there when he showed up.

Bart told us he was a talent booking agent driving a cab temporarily; you know how the business goes. He said he'd traveled around the world, worked in Puerto Rico as a bartender, an editor and a maitre d' at a kosher hotel, and then a record producer. Agents say wild stuff, we figured. Like claiming to work with us. That's why they're agents. We could use an agent, somebody to arrange gigs for us and handle the business end of things. Bart had charmed his way in. He could have the job.

Like the rest of us, Bart's real interest lay in making tapes not money. But by the time we noticed he wasn't putting nearly the effort into the business he'd promised, he and Nancy had picked up where they'd left off, and nobody would try to break up a relationship that involved a member of the group. They chose two rooms at the end of the second floor hallway at Maple Tree Farm, next to the door leading to the bridge over the driveway and the unheated rooms in the back building. Along one side of the covered bridge was a line of shower stalls for the summer customers. No one used them now, and we had stuffed them and a low-ceilinged bedroom above the rear apartment full of mattresses and bed frames we couldn't use on the assumption that was one room we'd never need to use.

∞∞∞∞∞∞∞

Chuck had met Ann in the winter of 1970 at the *Vision and Television* show at Brandeis University's Rose Art Gallery, one of the first major museum presentations of video art. Annie was a former Brandeis student working at the museum, and she graciously found room for some of us to crash on the floor of her tiny, off-campus apartment. We still thought of ourselves as undiscovered stars–so undiscovered we couldn't afford a motel and yet so big we swaggered into that show like we were the hottest thing in town. It didn't take long to figure out the fourteen video artists from around the Northeast, whose work was also featured in the show, didn't think as highly of us as we thought of ourselves.

David, never one to disappoint a crowd, managed to tumble backwards in his chair, falling into a fountain in the midst of a panel discussion we were televising via closed circuit throughout the museum. He was holding a video camera when it happened and got the whole spill on tape. Nam June gleefully applauded his clumsiness as spontaneous video art. Other panelists smiled wanly or shook their heads, suspicious Videofreex had once again engineered a way to make themselves the center of attention. For me, sitting in the makeshift pen that served as our open-to-the-world control room, David's pratfall was a welcome break from the insufferable tedium of artists talking about the meaning of their work.

Video was a novelty then that only occasionally attracted the attention of museums and galleries where big money could be found. I had no idea how to woo rich patrons, nor even that such a courtship was desirable. Meanwhile plenty of artists were willing to point out that our documentary past made our work–sniff–rather a bit suspect. After all, could you really call it art? So, here we were, faced with that same question: What were we? Buffoons? Commercial interlopers? Techno-hippies? Had we had a show in a gallery or museum that really mattered?

Davidson and David felt acutely the snubs of art world snobbery and weren't above a bit of it themselves. They understood that promoting your own work required a readiness to squelch competitors when the strategic opportunity arose. Carol and I, on the other hand, blithely dismissed the whole scene as bloodless self-indulgence. Others in the group fell somewhere in the middle, depending on the circumstances.

We managed to skirt these divergent sensibilities even after we moved to Lanesville. The giddy ride with Don West and the pressures of trying to make it in the city despite the odds against us had forged strong bonds among the Videofreex. But over time these essential differences in our aesthetic priorities resurfaced, and the gulfs between us widened.

After Vision and Television, Annie and Chuck carried on a long distance relationship for a few months until she consented to move to our Prince Street loft in Soho. Davidson already lived there in a tent-like structure of black polyethylene he draped from the sprinkler pipes in the studio. Chuck and Ann slept on a pull-out couch in Chuck's shop. The shop was taller than it was wide. Its grimy windows looked out on the gray brick of adjacent lofts. To reach the shop they had to walk through the editing and control room, where people often worked late into the night. For Ann and Chuck, Lanesville meant a real step up to luxury. They chose two rooms at the top of the stairway, right above the editing/control room, where the sounds of late night sessions wafted into their bedroom much as they must have penetrated the shop walls at Prince Street.

David and Oberon had a sunny room with bay windows on the second floor above the kitchen. He painted the floor white, mounted a leafless branch against one wall and let Oberon fly free. The floor became a patchwork of guano, but David didn't seem to notice. He was gone frequently, but Oberon befriended no one but David. He'd move close as if to accept my offer of a peanut, and then he'd bite me hard on the finger.

Curtis, Davidson and Frances joined Skip on the third floor. Sam still kept a couple of small rooms there, where he stored his belongings behind a door secured by a big padlock. After considerable cajoling, Davidson convinced him to clear out one of these rooms so he could build a darkroom.

Carol and I had two rooms on the second floor, which looked out over Route 214 and the Stony Clove valley toward Chichester and Phoenicia. In the spring and summer the sun filtered through the tall maples along the ledge above the highway and surrounding the house. In the winter the wind bellowed through the clove and shook the whole house. Cold air snuck through rattling windows no matter how tightly we shut them.

∞∞∞∞∞∞

Artists and producers from around the world visited us in Lanesville, some to work, some to socialize and many to do both. Few stayed longer than a week or so at a time, just long enough to edit or refine their tapes, to shoot something with one of us, to plan a cooperative production, to share ideas. Only one or two were invited to stay, and then not for long. We asked a handful of people to leave, usually for transgressions not easily defined. The first was a friend of David's named Bob Quinn.

Bob had a company that put on light shows for rock concerts at places like the Fillmore East and the Electric Circus, two of the main rock music halls of the '60s. Both places had closed by the time we got to Lanesville, and the hot new rock groups were looking for something more than the large screen projections of the amorphous shapes and the oozing colors that provided the signature trippy background for so many acid bands. Bob and David had spent hours cloaked in a ganja haze hatching all sorts of fantastic multi-media schemes.

About a week after we moved to the farm, Bob arrived accompanied by his wife and daughter, a toddler. During dinner that night, I saw Bob and David carrying one of the extra beds from upstairs across the music room toward the kitchen. I asked where they planned to store it, and David said in the back where Bob and his family would be living. The food dropped out of my mouth. David had said nothing to me about someone new joining the group. Did any else know about it? Murmurs around the table. No. Uh uh. Not me. We'd never discussed precisely how somebody could establish "membership," but it sure as hell didn't come down to a unilateral decision if it wasn't conjugal.

They continued on their way with the bed while the rest of us had a quiet dinner table conference. We'd struggled through lean times and now David wanted to spread our resources thinner. No way. None of us felt comfortable having a child around, either. This wasn't any place for kids, we agreed. We had work to do. Bob and family could visit, not stay. I was appointed to tell him because I'd say it most diplomatically and because I'd opened my mouth about it in the first place. Davidson volunteered to join me, feeling guilty because he'd been party to some of the hazy plans. We cornered David alone. At first he couldn't see what he'd done wrong. Share the wealth, new creative energy, a family scene ... wheeze, wheeze, wheeze. It took the two of us to wear him down. As a delegation, we gave the news to Bob. We're doing you a favor, really. David looked on sheepishly. Bob shrugged. Bad trip. He never returned to the farm and I never saw him or his light shows again.

∞∞∞∞∞∞

We had so many guests show up the first few weeks after we arrived at the farm, we never knew how many people would be eating dinner until everybody sat down at the table on the porch. Sam compounded the natural disorientation that accompanies moving to a new home by creating a minor turmoil every morning. He couldn't restrain his curiosity. He wanted to help, and I think he found our energy infectious. We were building things, and the sound of saws and hammers was music to his ears. Rooms rapidly changed character. Did he care if we ripped out this fixture? "I don't care as long as you're not tearing mine house down." But it was his house and he did care. "Wai' a minute. What you're doing here?"

Even when we weren't working, Sam came by. Some of us were lying around the side lawn talking with him on our first Fourth of July at the farm, when one of our guests, a spacy young woman named Laurel, appeared on the lawn wearing nothing but a hip-length, leather fringe vest and moccasins. She announced she was going hiking in the woods and was looking for someone to accompany her. Idly, she asked Sam if he'd like to go. He took off his cap, scratched the white stubble on the side of his head and made a low, growly sound. Laurel had no trouble finding another hiking companion.

Like Laurel, we lacked a certain sensitivity at first to the protocols of life in Lanesville. Chuck, Davidson and I worked late into the night and started again early in the morning those first weeks building the new fixtures we needed for our equipment. We made a racket that sounded perfectly normal by city standards. A few days later, the elderly people who summered in the house next door, the place Louise Doyle had shown us when we first came to Lanesville, put a For Sale sign on their front lawn. They moved out before the end of August.

We must have committed other faux pas we never knew of and might not have cared to correct if we had. We could not, for instance, have endeared ourselves to a succession of properly conservative postmistresses for the weekly copies of *Granma*, the official propaganda organ of the Cuban government, which we received thanks to Noah having put us on their mailing list. Nor could they have been pleased about our monthly copy of the colorful and glossy *Soviet Life*, a ridiculously saccharine portrayal of the USSR under Brezhnev. Sam was thrilled with this magazine, and we gave him every copy we got, usually without opening it.

One day a couple of years after we had moved in, we got a panicked call demanding that Ann come to the post office window immediately. Her package had arrived and was making ominous noises. She had ordered ladybugs as a natural insecticide for the garden and they were quite literally bouncing off the walls of the box. The postmistress detested crawly things and wanted them out of her cubicle before they burst loose and attacked her. She insisted Annie get her bugs by some other means in the future.

We never met the elderly folks we routed from the house next door. And the house on the far side of them was vacant when we moved in. These were the only dwellings close to Maple Tree Farm, and that, to us, was one of the charms of our new home. None of us felt any urgency about getting to know the neighbors. Even if we had wanted to, we wouldn't have known where to turn. Not one of the six of us reared in Protestant homes would succumb to our roots and attend the Lanesville Methodist Church. Yet the church, above the highway just south of the general store, was the only traditional organization we were aware of in the whole settlement. Not even a fire house here. Gert and Gene came by to pick up the garbage, but they seldom stopped to talk. There was Sam, and we made forays to the general store and to Doyle's bar, run by Louise's husband, Jerry. We walked in the woods, an unlikely place to meet anyone. And we traveled: to the city to work or to touch base with the video scene, and around upstate in our evangelical crusade to bring the wonders of video to the masses. We were looking outward from Lanesville, and bringing our own friends in, almost as if we were second-home residents, tourists or passers-by. Lanesville was our place of residence, not our community.

5
Porcupine Problems

Sam missed the wedding of Cy and Curtis. A few days before, he had walked in the front door unannounced. That in itself was unusual because of his habit of stepping inside the kitchen door and then asking permission to enter. He went directly to the free-standing closet in the front hall and pulled out his gray suit. "Mine son, Leo, is very, very ill," he said. He was leaving for Brooklyn right away. Eddie, who ran the body shop next door to him, was giving him and Miriam a lift to the bus.

Two days later Leo died. Sam and Miriam stayed in the city a while, and when they returned Sam didn't come over for several days. During this period I passed their house and saw them through the window, sitting at their kitchen table.

Sam came to the door. It was dusk. He looked confused. His eyes were red. "Oh, Meester Parry, I don't recognize you." He insisted I come in. We sat around the table. It had a speckled formica top with a metal rim, and was bare except for salt and pepper sellers and Miriam's sugar substitute. They offered me a cup of tea and a plain cookie from a box in the cupboard. Leo was their only child. The doctors thought maybe it had been equine encephalitis–there'd been a big scare about the mosquito-borne disease over the summer–but no, probably a stroke, and at only thirty-nine, with a wife and three children. "I knew," said Miriam. "I knew when they called they wouldn't tell me he's dead, but a mother knows." It grew dark outside. Sam's sister and brother-in-law, Bertha and Joe Keley, had been by earlier, and Eddie and his wife, Edna. Other than that, they were alone in Lanesville.

∞∞∞∞∞∞

Around the time of the First World War, a writer named T. Morris Longstreth, who wandered the Catskills, had singled out Lanesville in a book about his journeys, as a "friendly village" in a setting so peaceful as to be "susceptible only to little ills which a neighbor's sympathy could soothe." Half century earlier, a devoutly religious Lanesville resident, Sobrina Williams, extoled the virtues of this tight-knit community, then called Bushkill. She wrote of taking chicken soup to a family stricken with typhoid living in a house with "the windows broken out, and rags in the place of glass."

Once Lanesville had subsistence farming and woodworking factories, ice houses and small quarries. The railroad came in 1882, and from then through the first half of this century, business in Lanesville meant boarding houses. The settlement didn't have electricity until the New Deal in the early 1930s, when crews first laid pavement over the dirt road called Route 214. In 1940 the railroad, which brought what little prosperity there was, rolled up its tracks as scrap iron for a war about to happen. Improvements to the road helped cushion the impact of the loss of the railroad, encouraging some tourists to come by car; but always there were fewer boarders than the year before, and after the war it's hard to say what kept the community together.

We settled in Lanesville because we didn't have the money to live any place we wanted and because serendipity had led us to Maple Tree Farm. In calmer moments between our journeys I had begun to wonder what kind of place we had picked. Did the silence that greeted the death of Leo Ginsberg, the absence of traditional expressions of neighbors' sympathy, the containers of food, the flowers, cards, did this void of compassion provide a clue? I couldn't decipher it if it did. And like the rest of the Videofreex, I hadn't even concluded we should bother to understand the community. We might never have made the effort, either, if it hadn't been for Joseph Paul.

∞∞∞∞∞∞

Joseph Paul Ferraro was born and raised in Yonkers, a gray, working-class city now indistinguishably extruded from the Bronx immediately to its south. His mother, a hair dresser, owned a modest mock-Tudor home on a cliff above the riverside railroad tracks. As a kid in the 1960s, he used to stare across the Hudson, looking over the stone face of the Palisades toward the Armstrong tower in Alpine. The tower is a massive steel grid that rises over 300 feet into the Jersey sky. Its arms sprout radio antennas like cilia on a sea creature.

The man who built that tower, Edwin Howard Armstrong, was also a native of Yonkers. Working at Columbia University in the 1930s, he'd grown dissatisfied with the quality of commercial radio transmissions. This was the golden age of radio when the medium worked on the principle of amplitude modulation, or AM, and much of the audience knew nothing other than staticky, whiny, erratic reception. Armstrong invented a new way to broadcast radio waves that did not inherently distort the voice or the music being transmitted. He called his invention frequency modulation, or FM.

Armstrong possessed a genius for engineering, not business, and the businessmen at RCA, then the largest of the radio networks, decided to kill the FM idea. They had a huge investment in AM and no interest in fostering a competing radio service. Their next business venture was already under development and had nothing to do with improving radio. They planned to put their resources into a new technology called television. They viewed Armstrong's invention as a threat, not an opportunity, and they knew just what to do about it. Not long before, they had successfully muscled Philo T. Farnsworth out of his invention of the electronic scanning system used to create television images. No big deal, then, for RCA lawyers to slap a legal hammerlock on Armstrong and his invention. In 1953, exhausted after years of court battles with RCA, Armstrong committed suicide. He jumped out a window.

Joseph Paul wanted to climb the Alpine tower, to explore its arms and to mount an antenna of his own, like a mountain climber conquering Everest. But he didn't have the physical attributes to pull off a stunt like that. Smart, arrogant, flaccid, obsessed by technology and its unconventional, mischievous potential, he had studied Armstrong's tragic story, and it cemented his distaste for corporate media.

Joseph Paul had one close friend, Allan Weiner, who shared his fascination with radio technology. Together, Allan and Joseph Paul figured out how to modify military surplus radio gear available by the carload on Canal Street in lower Manhattan. In their basement workshops, they cobbled together AM and FM transmitters whose signals could be received on ordinary radios miles away. Instead of the Armstrong tower, J.P. used the roof of his mother's house. Allan had companion AM and FM stations at his home.

While they were still in high school, Joseph Paul and Allan went on the air with their home-built transmitters, alternating nights, so that one evening the signal would come from J.P.'s basement, the next from Allan's. They played whatever they wanted, said what they wished and invited friends over to share in the fun. Neither one of them had a license from the Federal Communications Commission to broadcast; both stations were totally illegal.

J.P. and Allan set out to challenge commercial radio, two fleas in a headlong assault on the elephant. But their Falling Star Radio Network caused no immediate uproar when they went on the air in 1971. Instead, they broadcast for months, slowly building a local audience for their mix of rock music and anti-war, anti-power-structure patter.

Thanks in part to RCA, the few FM stations that had ventured onto the airwaves in the 1950s and early '60s languished for a lack of listeners with radios to receive them. But by the time the two Yonkers pirates turned on their transmitters, the situation had changed radically. Baby boomers had arrived as a market force, and nowhere was the economic power of the generation more unmistakable than in the music business. That the Woodstock Festival could, with modest advertising, draw half a million people, was only a small indication of what was happening. Record companies scrambled to take advantage of acid rock albums that intentionally defied the limits of time and censorship imposed by top 40 play lists on AM radio. Suddenly, FM, with its higher quality sound, became the medium of the generation. First in San Francisco and shortly after in other markets, FM stations playing rock became profitable, wildly profitable, so profitable that when Joseph Paul and Allan went on the air, the big New York stations didn't notice or couldn't be bothered.

The local ham radio operators did notice, and they were pissed. Hams take their licenses from the FCC very seriously. Amateur radio operators see unlicensed interlopers, who unintentionally or otherwise clog frequencies and disturb their worldwide chatter, as agents of the type of airwave anarchy that prevailed at the outset of commercial radio shortly after the First World War. It took government intervention to straighten out that mess. The Yonkers hams found allies in outrage in the person of local scoutmasters, some of whose charges probably were working on their radio merit badges. Having two pirate radio stations operating in their back yard became morally intolerable to the scoutmasters–an especially bad example for impressionable young scouts and would-be hams.

At first they warned J.P. and Allan. Do the right thing, they said. Get a license; adhere to the law.

A license was out of the question. It would have been too expensive to pursue even if a frequency had been available, which it wasn't, and if J.P. and Allan had wanted to get one, which they didn't. They scorned the admonishments. They taunted their antagonists on the air. Tools of the establishment. Straights. Reactionaries. Well, maybe the scoutmasters and hams couldn't do much about the moral decay of society as a whole, but they had to draw the line somewhere. They reported the two pirates to the FCC.

The FCC's Field Operations Bureau took its time investigating the case, eventually issuing a warning. No response other than more broadcasts. A second, official, sterner warning. Still the nightly broadcasts. Bald-faced defiance. Finally, FCC officials and local cops swooped down on the desperados, clapping J.P. and Allan in handcuffs and impounding their equipment. The menace had been silenced ... for the moment.

The bust didn't worry J.P. or Allan too much. An arrest for political activities had become something of a rite of passage, and both young men, in their own self-indulgent ways, believed they had engaged in a fight for freedom of the airwaves. They had seized the initiative for free radio by seizing the airwaves in the world's largest media market. And besides, Allan's father was a well-to-do lawyer who could bail them out and defend them.

After protracted negotiations, Joseph Paul and Allan received suspended sentences and a warning to stay off the air or risk ending up behind bars. That meant Joseph Paul, who had no job, had plenty of time on his hands. He and Allan had become minor celebrities from the notoriety of their bust. Abbie Hoffman called. They met and he told them about us and our modulator. He said they should call us at Maple Tree Farm.

∞∞∞∞∞∞

J.P. sounded surprised I hadn't heard about his stations and his bust, a story he eagerly recounted over the phone. That's how out of touch we were, and I felt a little defensive about it. No *Post* that day, I guess. Did he know anything about TV broadcasting? He said he wasn't familiar with the specifics, but the general principles were similar to radio. I liked his confidence and the honesty of his answer. We had attracted plenty of kooky people, like the guy who got off the elevator at the Prince Street loft one day carrying a five-foot diameter metal disk, which he claimed held the secrets of the universe. Was there a machine in the studio he could play it on? I eased him back onto the elevator and told Morris the elevator operator we wouldn't be accepting any more deliveries that day. J.P. didn't strike me as that kind of weird, so I invited him to the farm.

He drove up in his mother's ancient Plymouth in the fall of 1971, not long after we'd moved in. He looked young, almost babyfaced. His deep, radio voice had masked his youth. I must have been three or four years his senior, but I felt worlds older and more experienced. He, Chuck and I spent the weekend talking about what we'd need to realize Abbie's dream of a pirate TV transmitter. The requirements had changed. No point in a signal that could blanket Manhattan when the mountains confined our audience to the Stony Clove valley.

Even though Chuck and Joseph Paul had many interests in common, I could sense a tension between them. We had come to believe that if Chuck couldn't fix a piece of electronic gear, a VTR or camera or whatever, it couldn't be fixed. We were proud of his accomplishments and overlooked his foibles, becoming almost possessive of him in public, as if having a technician of his caliber was an asset unique to Videofreex. At the beginning, it was. We had showered him with adulation for his gen-lock circuit, a black box he had designed and built, which allowed us to mix the video signals from a live camera and a pre-recorded videotape. For a long time, practically no one but broadcasters and Videofreex could gen-lock, and for a while we were the only ones doing it with 1/2" tape equipment. Chuck's gen-lock confirmed our position as the leading technical innovators among video groups. The other groups sniffed that they were more interested in content. We thought they were jealous of us and "our" Chuck. He worked obsessively, hunched over his workbench, chain smoking Lucky Strikes, which he lit with his soldering iron, slurping cups of coffee, hardly looking up for hours on end. It was as if Chuck entered the circuits he worked on, following the path of the electrons like an explorer. And if we placed too much of a burden on him in that we expected an unachievable string of gen-lock type successes, he did little to disabuse us of those expectations.

Unlike the rest of us, Chuck worked something like regular hours except in emergencies or during major productions that required his expertise. He shut down about five o'clock and drank as much as was available, which was usually no more than his half of a six-pack. He ate his dinner in front of the TV accompanied by Annie, and then went to bed, where he read science fiction novels by the carload. Most of our guests at the farm came because of what we produced, to learn how we did it or to share with us what they were doing, or simply to get away from the city for a few days. Unless they had some technical background, their interactions with Chuck usually consisted of introductions and a few words of admiration for his accomplishments and skill, after which Chuck simply leaned over his bench again. Chuck wasn't monkish in his habits, but the nature of his work was solitary, and the importance we attached to his skills lent him special status in the group. Now, out of nowhere came Joseph Paul, who claimed to know all about broadcast technology, the one thing we'd tried and failed to make work for us. I could see the arrival of Joseph Paul made Chuck uneasy.

J.P. didn't make things any smoother with his habit of chuckling a forced, "Oh, ho, Ho!" just before delivering a pronouncement on politics, science or art. He relied on this sort of distraction whenever Chuck probed the weaknesses of his technical knowledge. The relationship between them would have to be handled gingerly if we were to get on the air, and no one cared to take on that task but me.

I had some experience in this department. When I first met David, he occasionally spoke about a young video artist and self-taught engineer named Eric Siegel. Eric had attracted attention in 1968 at video show at the Midtown gallery of Howard Wise. His tape started with Einstein's head, which exploded into what, at the time, were fantastic shapes and colors. Eric moved to Sweden after that, but David convinced Don West to fly him back for *Subject to Change* on the promise that Eric would work magic: He would change our black and white videotapes into color. The public would not see colorized movies for almost a decade, and what Eric and a few other experimenters were doing–replacing variations of contrast and brightness in black and white video images with colors–sounded like pure science fiction.

I could sense Chuck getting a little edgy every time Eric's name was mentioned in reverential tones. And when Eric met us in California, I saw from the effete manner he used to dismiss Chuck that the best I could hope was that they wouldn't try to electrocute each other.

Working hastily and under difficult conditions, Eric concocted a circuit that produced some random colors, and a few of the effects were quite stunning at first. But the thrill wore off quickly, and Mike Dann and his CBS cadre had no illusion they were watching color tapes. The colorizer components proved unreliable and the few times we tried to use it, the device muddied the video images beneath the colors to such a degree it seemed to defeat the purpose. Eric perfected later models and went on to make other devices that improved the quality of videotapes made on low cost VTRs. But for all his fine ideas, he never seemed able to cash in on what came of them. He was a loner and made it clear in a standoffish way he had no interest in joining Videofreex. I reminded Chuck of this whenever he growled about people comparing him unfavorably with Eric. When Don's show died, Eric struck out on his own. Chuck never gloated at the dismal performance of the Siegel colorizer, but I know he was glad to see Eric go.

J.P. knew enough about TV broadcasting to understand he could not easily adapt the types of army surplus amplifiers he and Allan had used for their stations in Yonkers for use as pirate TV transmitters. Perhaps something would turn up, he said. Otherwise, building a transmitter would be an expensive and complex process, one J.P. reluctantly admitted he wasn't qualified to do. That gave Chuck his opening. He confidently predicted he could whip up a mini-transmitter using a $50 transistor we'd found through *Electronic Design* magazine. His announcement startled me because he had always put off broadcast projects before, and he'd taken heat in the group because so many of his promises, including the gen-lock, took much longer than predicted to fulfill. But I was not about to discourage him. Even though our money from the first state grant was running out, we could buy the transistor if we scrimped somewhere else.

Having thus reasserted his preeminent position, Chuck agreed J.P. should scrounge around Canal Street in Manhattan for the pieces needed to build us an antenna suitable for broadcasting to the Stony Clove valley on Channel 3. Chuck had never built an antenna and seemed disinclined to learn how, given J.P.'s clear head start in that field. J.P. said he'd need a few months to come up with an appropriate antenna, a timetable that got no argument from Chuck.

∞∞∞∞∞

No one pressed me for an update on a Videofreex broadcasting system, although I frequently queried Chuck as to the progress he had made on the transmitter. Keeping J.P.'s encounter with the FCC in mind, as well as our own failure in Washington, Chuck and I tacitly decided to downplay the project for the time being.

Even if we had wanted to make broadcasting a higher group priority, I can't imagine how we could have during that fall. Video projects–the workshops we'd committed to perform for the grant, as well as our own tapes–consumed most of our time. We frequently traveled to Rochester, where we'd linked up with a film maker named Bonnie Kline, who had previously done pioneering work with the National Film Board of Canada. Together, we recruited and trained a new video group called Portable Channel, whose members had begun making their own community programming.

I also commuted weekly to Manhattan to teach video production to junior high school kids at the George Washington High School Annex in a tough section of Washington Heights. More than once I had to hitchhike from Lanesville to the city, lugging a box of video accessories with me, because all our vehicles were already on the road. But any minor hardships were offset for me by the excitement of those students, most of them Hispanic, African-American and Asian-American, as they made tapes that reflected their concerns in a language and style they found comfortable.

They played roles, they told their stories directly to the camera, they developed the tapes from concept through production and editing, and then played them for their classmates. Despite inadequate budgets, dingy classrooms filled with noise, unreliable equipment and the taunting of peers, these kids managed to taste the creative process. They learned how to work as a group, and the class afforded them an opportunity to exercise a small measure of power over their own images and how others perceived them. I wish I knew whether the experience affected their lives. All I can say for sure is that their teacher told me she had never seen them so focused and unguardedly enthusiastic.

I found that class exhausting and intoxicating, and from all the contacts we had in the expanding video movement I knew my experience differed little from innumerable variations of the same project throughout the state and around the country. And yet even as so much energy went into these workshops, none of us gave much thought to how we might consolidate the changes we were so certain we'd wrought. We never discussed methods of measuring our impact, nor of setting benchmarks for the future. The furious pace of our work overwhelmed any sentiment for reflection. We wanted no distance from the revolution we were making.

Eager to fulfill the terms of our grant and ensure the likelihood of refunding, Videofreex stepped up our shuttling back and forth to Rochester, working with Portable

Channel and giving video as high a profile as possible. Carol, Nancy, Curtis and Frances taped an upstate Women's Liberation Conference. We tried to figure out how video could help efforts to restore an 1882 opera house in the old canal hamlet of Earlville. Sometimes we'd happen on events, like the Purple Lancers. Lanesville became a haven, a rest stop from life on the road.

We had so much to do at first we hardly had time for internal squabbles. The old sore points of access to equipment subsided as we worked collectively on Council projects. Annie maintained her distance from production by assuming a measure of control over the shopping and cooking. The aggressive way the rest of us pursued our projects meant we didn't stop to ask whether she wanted to be included. She didn't demand her place in the video work so she didn't get it.

Frances didn't seem to want a place in the video pecking order, either. This had nothing to do with the force of her personality. She had a deep voice and wore large, tinted glasses, and she possessed all the outward strength she would have needed to become a tapemaker like the rest of us. She was a poet, more comfortable with words than images, and her contribution to the group in those early, unsettled months was to produce a newsletter, *Maple Tree Farm Report*, which we mailed to all the video people we knew. The first issue was a simple three-page, typewritten flier that appeared late that summer, and to our surprise, people wrote to us from all over the U.S. and Canada asking to be placed on the mailing list. It was our first inkling of how rapidly the video movement was growing.

The second issue was ready by mid-fall. Buoyed by the success of the first report, it had grown to a tabloid sized publication, contained illustrations by Annie and a feminist cartoon by Frances featuring a character called Video Velvet, and the whole thing was hand lettered by a friend of Davidson's, a cartoonist named Willie Murphy. When his hands gave out from the sheer volume of scribe work, Videofreex pitched in. Once again the response was positive. But this second issue of the *Maple Tree Farm Report* proved to be the last. Frances left the group a few months later and no one else wanted to carry on her project. A newsletter took too much time and effort, and most of us found we preferred making video to writing about it.

Not one of our workshops or showings outside the farm that summer and fall involved the people of Lanesville. Not a single copy of that modest newsletter was mailed to a neighbor. We hardly knew anybody in Lanesville other than Sam and Miriam and the couple who lived in the bungalow behind Maple Tree Farm. The report was aimed at the Council staff as much as anyone. None of us considered that anyone in Lanesville would want to read about what we were doing. Why would anyone care?

The newsletter did draw considerable outside attention to our work at Maple Tree Farm, now simply called "the farm." It legitimized our work in a way the small group showings of our tapes could not. But Frances' determination to work with words rather than video reinforced her position as an outsider, an observer rather than a participant.

By fall, we could see she and Davidson had decoupled, although they issued no formal announcement. That made other women in the group mistrustful. After the departure of Curtis, all of them but Frances lived monogamously. Their suspicions were subtle and seldom voiced, an undercurrent of wariness.

Some time after Curtis's wedding, I trooped up to the third floor to hand in my submission for the second edition of the newsletter. Frances wasn't in her room. "I'm in the bath," she called out. I told her I'd leave my manuscript in her room. "No, give it to me, the door's unlocked." I hesitated, then pushed the door open slightly. Frances was stretched out in the old fashioned claw-foot tub, with no bubbles or towel to disguise her long, shapely body. I said nothing. Several seconds went by in silence. "You want to give that to me?" She was looking at the papers in my hand. I almost dropped them on the floor as I reached over from the threshold. I didn't want to get too close. That was fine with her. "Close the door on the way out," she said. I did, finding myself in the dark hallway, a little dizzy.

A year earlier, long before Lanesville and the appearance of Frances, Videofreex had rented a summer house in the Delaware Gap in Pennsylvania. One particularly hot day, all us had gone swimming nude in the Delaware River. Power boaters hovered in the channel and nearly ran aground ogling us naked hippies. There had been the body painting party two years earlier, too, which Nancy and Carol had declined to attend. And we all knew tidbits of the amorous adventures of David and Skip, and occasionally Davidson now that he was single, offhandedly reported or, better, discovered on tapes never intended for public viewing. Those incidents only enhanced our sense of shared intimacy, of family secrets and a collective understanding of the bounds of decorum. Frances' behavior fit within our long tradition of exhibitionism, not seduction. Still, my surprise and what I took to be the calculated way she'd exposed herself shook me. I saw it as a breach of group etiquette. As much as turning me on, it made me wary, too, but I kept my misgivings to myself.

∞∞∞∞∞∞

Winter began that year with an unexpected Thanksgiving Day snowstorm. Fourteen inches of wet snow to greet Valeska and her German TV crew. The plows had all they could do to keep the road through the notch open so the skiers could pass. To the south, the steep curves of Sunshine Hill became nearly impassable unless you waited at the bottom for the plow to come along and followed it up the hill. The departure of the warm weather also meant a slight decrease in the number of guests traveling to Maple Tree Farm. David, Skip, Carol and I made plans to travel that winter. In the meantime, a sense of isolation set in. We all wanted some contact with the outside world. For many of us, that meant TV from New York City. Receiving TV rather than broadcasting it became a priority.

In the midst of another snowstorm, we set a 50-foot mast in the ground in a small clearing more than half a mile up the mountainside. The plot had a commanding view of the valley below. On a clear day, you could look over the tops of the low peaks of the southern Catskills to the Shawangunk Ridge and south from there to the western rim of the Hudson Valley. That southerly view was exactly what we needed, because from that point, we could pick up the VHF channels (2-13) from the city. To get the signal from the clearing down to the house, we had to drape two sets of cables through the trees–coaxial cable for the TV signal and a power line to run some used cable TV amplifiers we'd picked up. We housed one of the amplifiers in a large mailbox nailed to a tree midway along the route to the clearing. At the top, next to the antenna, Davidson built a what looked like a miniature outhouse to hold the second amplifier and other equipment we needed to process the signals.

Excitement mixed with congratulations the day we sat around the TV in the viewing room and watched the news from New York City. We had reconnected with the metropolis. Sam was astounded and asked whether we could get the signal to him. Sure, no problem. We ran a cable to his house, and later to the new neighbors next door. Slowly, word got around. There were cautious inquiries, usually through Sam. Could we extend the service? What would it cost to share our signal? Without planning it that way, and with no authorization from the town, it seemed as if we might be getting into the pirate cable TV business. There were so few homes in Lanesville, no one else was likely to offer the service.

Despite our initial euphoria, the quality of reception from the city varied greatly day to day, sometimes hour to hour, and not always stemming from the varieties of electronic interference that can plague TV transmissions. A few weeks after we erected the antenna, I was walking with Mushroom along one of the old logging trails and quarry roads that crisscrossed the mountainside behind the house. We were not far from the antenna clearing when he bolted up the mountain. I called to him, unable to see what he was after. He kept going, ignoring my commands. I looked upslope of him and saw a large, round form lumbering uphill. It was gray, about four feet high and it swayed as it moved. I thought it must be a bear. I screamed at Mushroom to come back, but by then, he had attacked the animal with a silent fury I'd never seen. I heard him squeal and thought he was a goner. Worse, I could be next. I froze, unsure whether to run for safety or try to help him. At that moment, I saw Mushroom loping back in my direction apparently unharmed. I felt relieved to see him safe, but not all certain the bear or one of its companions wasn't lurking nearby.

Mushroom stopped a few times to scratch. Remarkably casual of him, I thought. When he got to me, I saw why. It hadn't been a bear. His whole face, his mouth and tongue were studded with porcupine quills. He looked like a walking hors d'oeuvres tray. He was miserable and confused. I turned and ran down the mountain, urging him to follow. He didn't need much encouragement. He frothed at the mouth and every so often he stopped and tried unsuccessfully to dislodge the quills with his paws. At the farm, we trundled him into the car and drove up to Windham, twenty miles away, to the vet.

Less than two weeks later, we were back at the same vet, with Mushroom once again knocked out by anaesthetic while the vet and I used surgical clamps to remove the quills. The vet said some dogs never learned about porcupines, and Mushroom might be one of them. That turned out to be the case.

Meanwhile, we had another porcupine problem. They were eating our TV reception from New York. There seemed no part of the system they didn't find tasty. One trip up the mountain, we discovered several of the arms of our antenna gnawed off. Then they went after the outhouse and ate most of it before we covered it in steel mesh. After that, they turned to the one thing we could not protect: the cables. They ate everything, even the half-inch thick, aluminum-skinned antenna cable. They liked the power cable best and seemed to find it most delectable when current was running through it. You could tell when the porcupines were supping. The signal would go on and off as they chewed through the power line. In the end it went off for good; the porcupines won, and we brought all the equipment off the mountain.

By then, Chuck had discovered we could pull in the signals from two plodding UHF stations in the coal mining towns of Scranton and Wilkes-Barre, Pennsylvania. That happened by a fluke of topography, saddles in just the right spots of the intervening mountains. In his home in an old bread truck parked in the field behind Gert's house, Gene could receive a station from Utica. Down below, at Gert's, the best picture usually came in from Hartford, Connecticut. The rest of our neighbors pulled in Albany or nothing.

∞∞∞∞∞∞

The porcupines managed to destroy our cable system and Mushroom, too. Because he couldn't stay inside for prolonged periods, we had to keep him on a chain, which seemed heartless to me. Whenever he got off the chain, he headed straight up the mountain and frequently returned with a face full of quills. He was defending our cable system and couldn't understand there was nothing left to defend. Once we got a call from people who'd found him on the other side of Hunter Mountain in Spruceton, a community so remote and small it made Lanesville look like a metropolis. The next time he ran away, no one called. He never returned. Gene's denial that he'd eaten him was the last we ever heard of Mushroom.

6
Something Extounding

As the snow and ice closed in after Thanksgiving 1971, Maple Tree Farm, which had seemed so huge when we'd arrived a few months before, began to feel much smaller, and the habits of other Videofreex, particularly the bad habits, became far more noticeable. The prospect of spending the entire winter cooped up together in Lanesville served as a powerful motivation to make travel plans.

David, as usual, went first. In the fall, he had taped gatherings in Washington, D.C., of Jewish mystics and scholars. Having established his bonafides as a video observer of Jewish life, he and Allon Schoener arranged a personal grant for David to travel to Israel to shoot tapes for Hadassah, the Jewish women's organization. Schoener, a veteran of art world politics, was an old co-conspirator of David's, who had frequently helped us craft applications that the Council and other agencies had funded. This time, though, the money wasn't intended for a group project, and David and Allon would have complete control. David expected to take a Videofreex portapak with him and to buy and use at least a dozen blank, half-hour tapes–an extraordinary luxury by our frugal standards–contributing neither tape nor money to the general Media Bus account. We were finding it hard to meet expenses and pay ourselves our $25 per week salaries, and we had just learned the Council planned to reduce our grant next year. His timing was terrible.

David earnestly explained how tapes from Israel would enhance the scope of our library. Wheeze, wheeze. And when the howls that greeted this self-serving line died down, he reluctantly agreed to share a small part of the grant. The tapes he came home with pleased Hadassah–a tour of the organization's hospital, an interview with Teddy Kollek, the mayor of Jerusalem, etc., and a visit with a group of radical Jews who called themselves Black Panthers. They played on multiple monitors as a major show at the Jewish Museum in New York City, "The Word From Jerusalem," a celebration of Hadassah's sixtieth anniversary. It rated front page coverage in the *Village Voice* newspaper, although the writer used the show as an excuse to recount her own experiences in that ancient city, leaving readers with relatively little description of the tapes themselves. I recall only one sequence that struck me as memorable, a disturbing tape shot in a tiled abattoir, where David followed a worker calmly slitting the throats of sheep one-by-one as he stands in the middle of the small herd.

VIDEOFREEX

The Jerusalem tapes ended up on our viewing room shelves along with hundreds of others neatly cataloged by Carol. Tapes with a catalog number and label could not be recycled without the tapemaker's permission. And occasionally, during one of our frequent tape shortages, someone would grumble about the untouchability of "the fucking Israel tapes." The story in the *Voice* was a triumph of sorts for the group identity. The writer referred only to Videofreex, naming Allon as the producer but never once mentioning David, who shot it all. That had to hurt, and the lingering bitterness over our Hadassah tapes squabble left David edgy about the ingratitude of his fellow Videofreex, and the rest of us no less uneasy about his private projects.

∞∞∞∞∞∞∞

In late March of '71, when we still lived and worked in the city, we received a letter from Algeria. It read, in part,

Comrades:

The International Section of the Black Panther Party has initiated a video tape program to be directed to the United States and Europe on a regular basis to cover the spectrum of the international anti-imperialist revolutionary movement ...

The letter requested a number of our tapes as listed in the Raindance group's broadsheet magazine *Radical Software*. The tapes included the segments of Fred Hampton, interviews with Abbie and Jerry Rubin, edits from a number of street demonstrations, a laconic piece David had shot from a sailboat that had tacked by Sing Sing prison as it sailed up the Hudson, and, lastly a series I was working on called *Dr. Electron*, which included a how-to segment on soldering connectors. The letter was signed, "ALL POWER TO THE PEOPLE, Eldridge Cleaver, Minister of Information," and we were instructed to send the tapes to one Carole P. Roussopoulos, 18 Rue de l'Odeon, in Paris. We sent the tapes right away with a cover letter signed "Videofreex."

Carole and her husband, Paul, arrived at the farm that December. Neither was French, which probably explains why they comprised practically the entire video movement in France at the time. Carole was Swiss, a former model, who spoke very good, though heavily accented English. Paul was Greek and, we learned later, held a part-time job as a Ph.D. chemist with a major French pharmaceutical company. He spoke no English and she translated for him, as he chain-smoked Gauloises. They had shot tape together in the refugee camps of Jordan and covered French political demonstrations in addition to establishing regular contact with the real Black Panthers, or at least the faction that was in exile. Cleaver had narrated one of their tapes, and they knew Lily. I got the impression they didn't care for her. Carole and Paul had two children, a girl, Alexandre, a toddler, and an infant son, Geronimo. They'd left both of them in the care of the Panthers faction in Algeria before coming to the U.S.

Sony had marketed a European version of the portapak after introducing the product in the U.S. The French and the British each maintained two technical standards for

69

TV broadcasts, none of which were compatible with the U.S. standard. It looked as if exchanging videotapes was a technical impossibility. But Carole and Paul, calling themselves Video Out, had pressed ahead, pursuing a course similar to ours. They used a portapak to shoot their tapes and invested in a VTR using one-inch-wide tape to edit. They were the first video people from Europe we had met, and we talked excitedly for hours. Paul was particularly interested in what Joseph Paul and Chuck were up to. As independent producers, he and Carole had run into many of the same roadblocks at the government-controlled TV monopoly, ORTF, that U.S. networks had placed in our way. When Carole and Paul extended an offer for us to visit them in Paris, we accepted immediately.

Skip decided the time had come for him to take the trip to Europe he had postponed when he joined Videofreex. So he accompanied Carol to Nice, where Carol's parents had rented an apartment. Unlike David's trip, theirs was a vacation with no portapak to distract them. I stayed behind a few weeks to fulfill my promise to work on a major multimedia show of Cy Griffin's at the Everson Museum of Art in Syracuse, an institution determined to make a reputation for itself as a major player in the video and multi-media art scenes suddenly so trendy. Cy wanted me to handle all the technical stuff, and I went for days without sleep while I tried to turn his nebulous concepts into a reality that lit up screens and made noise. I discovered a long forgotten underground cable running between the museum and Syracuse University. I wrestled for days with endless technical and bureaucratic hassles so I could pipe an image into the show from a surveillance camera in a medical lab at the school, proving ... something, which I'm sure seemed quite important at the time. I collapsed from exhaustion just after the rest of the Freex arrived for the opening.

A few days later, the first week of February 1972, I boarded an Icelandic Airlines bargain basement, classless red-eye flight from New York to Reykjavik, where we were marched off the plane through the gift shop and back to our seats to meet the requirement that we had not taken a direct flight. Then on to Luxembourg. The train from Luxembourg to Paris was almost as expensive as the flight and about as unpleasant. The American woman in the seat across from me in the second class train compartment was determined to carry on a conversation: You seem like a perfectly nice person, she said, and I hope you'll represent our country well. Not like those protesters, that Abbie Hoffstein person.

Abbie's not a bad guy when you get to know him, I said. Her mouth twisted into an ugly little smile, and I got some sleep.

Carole and Paul had a sunny, comfortable, third-floor, walk-up apartment in an old building. A few doors away at the end of the street loomed the historic Odeon theater, though it was not nearly as imposing as the dragon-like concierge who did not let anyone or anything pass into or out of Number 18 that did not meet with her approval.

Carol met me at the apartment and for a few days we did touristy things, museums, Eiffel Tower, Champs Elysees. She was pregnant and vomited in some of the finest parts of the City of Light. We had no way to show the tapes I'd brought, and the language barrier quickly led to tedium as we sat through Video Out tapes of French demonstrations. The latest one Carole had shot covered a huge abortion rights demonstration in Paris. Abortion wasn't yet legal in France, nor would it be legal throughout the U.S. for another year, though the movement was building on both sides of the Atlantic. I tried to figure out how to beg off politely from the playback scheduled for a group of women gathering at the apartment. As I edged out, a tall, dignified woman with silvery hair entered the room. For our benefit, Carole introduced Simone de Beauvoir. She and the other French activists watched the tape without comment, becoming animated only in the lengthy discussion that followed. My Berlitz book French didn't allow me to follow much of the discourse, but from that encounter and scores of less memorable ones that followed, I left France believing that other than Carole and Paul, Parisians derived more pleasure from talking about video than they did from making tapes.

Because the apartment at 18 Rue de l'Odeon was the center of video activity in Paris, I suppose it's logical that Jack Moore would drop by. He called first, a lilting, breathless lisp. With Jack, there was always a reason, and this time he needed a portapak fixed and none of his phalanx of young male acolytes was up to the job, and after all, I was a member of the Videofreex, and Videofreex were techies, right? I couldn't make his VTR work and probably made it worse, but the attempt didn't hurt my reputation. So Jack bundled us into his miniature Citroen for the long, dismal ride to the Netherlands during which he rattled on the entire time, never repeating a story from his endless collection of personal fairy tales, all of which he swore were true.

A few years later I mentioned something about Jack Moore to Paul Roussopoulos. He looked at me and scowled. "Le pap du video," he grunted and went back to what he was doing. The video pope. He hadn't meant it as a compliment, but it wasn't a bad description. By his own estimation Jack was the central figure of European video. He had cooked up video schemes and scams all over the Continent. In most of Europe, television was a state monopoly, and bureaucrats and politicians saw the new video gear as a threat. They imposed registration requirements for owning VTRs, and they kept records of who bought blank tape. Jack had no use for convention, European or American. When, for instance, he found his video projector for a show in Switzerland had been waylaid at the border, he made a quick trip to Rome to meet with "the men in the large, dark hats," whose extraterritorial connections got the projector to its destination right on time. Jack had once designed opera sets; he loved theatricality, and doing business with the Mafia struck him as romantic rather than dangerous. And, he said, pursing his lips: They got us the proper papers to get the projector out again. Oh, they are one of the few truly international organizations!

Jack, owly gnome with a wild hedge of a beard, stood no more than a plump 5'3" tall, pulled his long hair back in a pony tail and had a wardrobe consisting exclusively of drab, blue coveralls and cowboy boots. His round face had surprisingly delicate features accentuated by wire-rim glasses that framed his small, quizzical eyes. He hailed from Gene Autry, Oklahoma. He was fluent in at least seven languages, and many of them sounded remarkably similar crossing his lips because they shared an uninflected intelligibility, which made him an original no matter what language he was speaking. He surrounded himself with young hippie men of various nationalities, and he possessed a showman's sense of what would shock but not outrage his audience. He was always on.

Jack lived on a houseboat in Amsterdam, or rather that was his base of operations. In the winter, the water was too cold, so he moved to a cramped apartment on the fringes of the flourishing–and officially unhassled–Amsterdam drug scene. He never stood still for long, always fussing over something, calling somebody somewhere far away, planning some huge scheme–video for the '72 Olympics in Munich, a show in London or Paris, a connection with Sony in Tokyo or the UN in New York. Hard to tell how much of it was real; all of it sounded at odds with his dingy surroundings. But you couldn't write him off because he made it seem as if nothing much would happen unless he made it happen. Jack knew the European video scene was years behind the U.S., what with governmental restrictions, international boundaries, technical disparities and animosities old and new. He saw us as a valuable ally in the U.S., a name to conjure with in Europe. Skip had joined us in Amsterdam, and Jack treated us all as visiting dignitaries–no less than what we had expected but more than we'd experienced in Paris. Skip stayed on after we left. David made his way from Israel to Amsterdam and met Skip and Jack there and went on to London. The grand tour of video.

∞∞∞∞∞∞

As we straggled back to Lanesville and compared notes, we understood for the first time that the video movement had become a truly international phenomenon. Formidable technical barriers still stood in the way of the free exchange of tapes between video makers in North America, Europe and Asia. But in many of the industrial countries of the non-communist world, individuals and groups had embraced portapak technology as a tool for social change or personal expression, or both. The abundance of money, technology and freedom put U.S. groups at the center of this new worldwide movement, and we saw ourselves at the center of the center.

That made readjusting to life at Maple Tree Farm all the harder. March presented us with an unremittingly bleak landscape. The naked hardwoods colored the mountainsides a necrotic purple-gray. The sun remained hidden above the clouds for weeks on end. Roadside slush formed permanent levees. The wind overturned metal chairs on the porch, a drainpipe sagged near the back entrance, drafts played along the kitchen floor.

The Council's cutback would mean even more parsimonious production limits and a stepped-up search for new sources of money. Most of the Videofreex had embarked on major projects of their own. I had begun to write a book on the care and feeding of portapaks, with a treatise on how they worked, all geared to the people who used this equipment, most of whom had little or no technical expertise. We all applied to the Council for funding as individual artists in addition to our application as a group. Except for Frances. She was gone.

I had been delegated by some of the others–not Davidson, who had new female companionship in the city, nor David, who had already departed for Israel–to tell her she didn't fit in as a member of the group and should leave. We sat across the kitchen table from each other the night before my flight to Europe. I think she already knew her will, no matter how strong, couldn't prevail over the collective decision. But she didn't trust that I represented the group. Maybe just Carol and I wanted her out. I had practiced this conversation in my mind and knew she would demand I list her transgressions so she could refute each one. I ticked them off, the most egregious of which had to do with her isolation from the rest of us, as if she boarded on the third floor, an anthropologist on a field trip in our midst. There had to have been something in there about her food preparation habits, too. But I went lightly over the picky stuff, not wanting to defend Chuck's etiquette, or mine.

David incited shouting matches with his selfishness and no one asked him to leave, she said.

I agreed. But he had standing as a founding member, and besides, whenever we all came down on David, he mended his ways for a while. She didn't seem willing to change, to adapt. She could have questioned whether we held her to a different standard, but she didn't.

Annie walked into the kitchen and got a glass of water. She said nothing at first. She and Chuck knew the confrontation was imminent, and now she had worked up the courage to come see where things stood. To my surprise, she joined in, telling Frances in a voice edged with anger: You know you don't fit in here. When Annie didn't return to the viewing room, Chuck ventured into the kitchen, ostensibly looking for her. The temperature of our words, which had risen for a moment, just as quickly subsided.

Without saying so, Frances had capitulated. She had no one to take her side and she knew it. Her jaw set. No tears showed in her eyes. Chuck and Ann left the room. Frances bargained for a comfortable exit. I wanted her gone by the time we returned from Europe. I said she could use the van to move her things.

∞∞∞∞∞∞∞∞

Carol had been furious with me for performing the marriage of Cy and Curtis while maintaining I didn't believe in marriage for myself. My lame excuse that the ceremony meant nothing more than a favor to some friends didn't wash. I had to admit the inconsistency. In October she told me she was pregnant. She would have an abortion if I said so, but the decision had to be mine. Otherwise, she wanted us to get married so the baby would be legitimate. None of this hippie naivete for her. She'd been a social worker in New York City. She knew the importance of a child having an identifiable father. We could get divorced right after the baby was born, she said. She stood in the doorway to the office off the kitchen crying. She rarely cried. If the issue had come down to whether we should try to have a baby, I would have said no in a minute. Too busy. And Maple Tree Farm: imagine bringing up a kid here! We didn't even like having kids visit. Always in the way, the little rats. No theoretical deliberation now, though. Abortion was legal in New York. It was her right, and she had delegated me to make the decision. All I had to do was give the word. I hesitated to do that, but if I didn't, how could I duck marriage? Forget the length of your hair–mine was nearly down to my butt by then–or your politics or your bell-bottom pants; this went to values, not appearances.

Carol and I got married in November in Tannersville at the former bowling alley made over into the town hall. Justice of the Peace Covey Chapman officiated. We'd had our blood test and filled out the license papers. We drove up on a chilly, overcast morning, with Carol worrying we'd arrive late and me reminding her that no one but Bart, Nancy and Skip planned to meet us there, and thinking of course that if we were late and the judge had left ....

I had forbidden them to tape the ceremony. I wanted no visual record. I had made my decision and would stick by it, but that didn't mean I had to surrender my perversity.

Chapman seemed slightly unnerved by the five of us. He led us into the town board meeting room and we all sat down at a table. Did we have anything particular we wanted him to say? "Just signing the paper would be fine," said Carol. Now that the deed was going to happen, she wanted it to go forward with a minimum of fuss.

"I once married a couple by a waterfall," Chapman encouraged us. He thought he should give us every chance to make the occasion memorable. I found it memorable enough already. Skip wandered to the back of the room to inspect the huge iron safe that held town valuables. Chapman eyed him nervously. "Well, I should read the basic vows," he said.

"You don't really have to," I volunteered.

"I think you should each at least say 'I do.'" He sounded disappointed.

"Okay, Okay," we agreed, and while Skip spun the tumbler on the safe, we exchanged the minimum vows.

"Wait a minute! You mean I missed it?" cried Skip in mock horror as we stood up. He kissed Bart. Carol kissed Nancy. Chapman shook his head and my hand and edged out of the room to give the certificate to the town clerk.

∞∞∞∞∞∞

All of this had transpired since we had last seen J.P. So when he showed up in March with the three-element yagi antenna, which looked like a contraption used to hang clothes up to dry in the basement, it took us a moment to refocus on the prospect of broadcasting.

Our newfound international perspective helped. For so long–too long–we had defined our work in terms of what we weren't: the networks, commercial, corporate, beholden to the government or any central economic power. We'd positioned ourselves wholeheartedly in the front lines of that anti-TV movement as soon as the CBS project fell through. These broad, bold negative statements covered up the lack of a clear definition of what it was we stood for. We got away with that while the movement was busy inventing itself, but in France and Germany and England, as well as in the U.S. and in Canada, where the National Film Board was pumping money into video projects, we had seen people reaching out to local audiences. All of them distanced themselves from the mass media without relying on we're-not-them distinctions, which had little practical value when it came down to winning the loyalty of the viewing public.

In our travels outside Lanesville we'd encountered a new phase of the video movement, a painfully slow effort by people using portapaks to find day-to-day ways of connecting with their communities through video. This movement didn't need a consistent ideology other than a shared dissatisfaction with the mass media's stranglehold on TV stations and cable systems. In the place of ideology we substituted a faith that once audiences learned how we could produce programming that reflected local concerns, they would demand to see that programming on their TV sets. The trick was to get to the audiences in the first place.

In Amsterdam, Jack Moore and his co-pranksters were showing videotapes through the hashish haze of a nightclub called Melkweg (Milky Way), and they'd developed a loyal following. In 1972, the Federal Communications Commission, prodded by activist Commissioner Nicholas Johnson, was just getting around to requiring cable TV systems with 3,500 or more subscribers to set aside channels for the public and local programmers to use. Less than ten percent of the nation's households had cable TV hook-ups, but local programmers from Appalachia to Washington state had already begun to tap into cable as an outlet for all sorts of community activities. This put us in an awkward position. We went around the world extoling the virtues of local programming, but we produced none for our own neighbors.

We knew if we waited for cable, we'd never have an audience. The nearest cable line was down the valley in Chichester, and the owner was having a hard enough time keeping the porcupines from eating his meager system serving Phoenicia and the immediate surrounding area. He had neither the money nor the required franchise from the town of Hunter to run cable across the county line to Lanesville. A local TV installer had applied to string TV cable for Hunter and Tannersville, but he, too, had no plans for Lanesville. The state constitution prohibits cables crossing the notch that cradles the highest point of Route 214 from Tannersville because the roadsides are forest preserve land, and more to the point, there simply weren't enough homes in Lanesville and Edgewood combined to make it profitable to build a separate system there. If we were going to program for our neighbors, the only way to reach them was over the airwaves.

Chuck had completed the amplifier, but he had no way to test it on the bench. His second-hand oscilloscope from Canal Street did not have the sensitivity to measure the frequencies that would indicate whether the circuits he'd built would work. We'd have to run an experiment in the real world to determine what, if anything, it could do. The heart of the amplifier, the $50 transistor, a flat disk about the size of a quarter, was mounted with its companion circuitry in a small aluminum box. To run it required another little box, a power supply, which Chuck purchased from an electronics supply store in Kingston. I had always envisioned a pirate transmitter as a kind of Dr. Frankenstein device, a Van de Graaff generator with bolts of static electricity sliding up and down gleaming posts. Ours looked a lot like a box of Animal Crackers. I consoled myself that Abbie would have approved if he could have seen it. He was not available. He had been busted in a cocaine raid, and maintaining his innocence and his inability to receive a fair trial, he had gone underground.

With our porcupine disaster in mind, we agreed that the Channel 3 antenna should be mounted on the peak of the roof at Maple Tree Farm rather than on the mountainside, where the signal from it might have reached more homes. We pointed it down the valley because most of the homes in Lanesville lay in that direction, and the valley curved sharply just north of the farm, making good reception unlikely in that direction no matter what we did.

We ran a cable from the first floor control room to the amplifier and, on the evening of March 18, 1972, a Saturday, we broadcast our first test signal from Lanesville TV. J.P., Nancy and Bart drove down the road to the home of one of the few neighbors we'd met. They called me from that home near the general store to say the reception was great. The next day, Sunday, we turned the transmitter on early, and pointed a camera at a test pattern with a note clipped to it announcing that we'd be on the air at 7 p.m. that day.

Down at Doyle's Bar, the picture of the pattern on the set above the bar wavered a bit, like a flag in the gentlest of breezes. Our test tone was accompanied by a hum of unknown origin. We weren't exactly regulars at Doyle's, but we'd kept up a cordial relationship since his wife had found the farm for us. Jerry Doyle had a red face, fat belly and a garrulous exterior, but he displayed his best Brooklyn–now Lanesville–ward captain demeanor as he grabbed the microphone from my hand and said to Carol, who was shooting tape for the first show, "This is going to be something new and something ex-tounding."

Driving along Route 214, we spotted Bruce, one of Gert's kids, a handsome, gangly adolescent. Like some of Gert's other offspring, Bruce dropped by the farm occasionally to observe, politely and shyly, what we were up to. Mostly I think, the kids didn't hang around because our work must have looked pretty boring after a while. Not this. Lanesville was getting its own TV station. What d'ya think about that, huh, Bruce? He kicked the snowbank a couple of times and barely looked up. "Okay, I guess," he replied slowly.

The first official, if you could call it that, broadcast of Lanesville TV, Channel 3, went on the air at 7 p.m. that night. It consisted of the short tapes we'd shot around town with Doyle and Bruce, and an old Bugs Bunny cartoon Bart had taped off the air in the city several years earlier. A number of us appeared on camera to introduce ourselves and our concept of a community TV station. Over and over we urged people to call us and describe the quality of their reception; and a surprising number of people did. I worked the switcher in the control room as well as appearing on camera. My microphone was an old one from the CBS project, and looked like a small gooseneck lamp. It kept getting caught in my hair.

On one of those early broadcasts, Jerry Doyle called to say his reception was perfect. Further down the road, Eva Quick said she could pick up either the picture or the sound, not both. At her neighbor's, though, both came in. In retrospect, none of this seems surprising. Our signal was erratic for a variety of reasons, not all of them within our control. The modulator was old and we had no way of tuning it adequately. Chuck's amplifier worked, but how well was open to question, and its performance varied from week to week, even minute to minute. Our internal communications system also left something to be desired. On broadcast days when everything was ready, someone would lean out of the control room on the first floor and shout up to Chuck on the third floor, "We're on the air!" That was his signal to turn on the transmitter.

The quality of the TV sets in Lanesville and Chichester didn't help, either. The often-cited statistic that Americans have more TV sets than bathtubs may well have held true even in the Stony Clove Valley, but because so many people scraped by below the poverty line, the sets were old and used the same types of cranky tubes that plagued the eldest of our gear. And if a set didn't work well, who would notice what with the reception so lousy at the best of times. Plenty of folks in Lanesville could have afforded a fancy new set, but they had no incentive to spend good money for fuzzy signals.

As we had learned with Sam's set, some homes got the best regular reception from Albany by pointing their antennas away from the Albany transmitters so they'd pick up a signal "bounce" from a mountainside, in essence, using the bowl of the mountains as a huge-and unpredictable-signal collection dish. This meant the antennas of many of the homes downhill from Maple Tree Farm also faced away from our transmitter, and because our signal bounce was extremely weak the alignment of antennas resulted in chronic variations in the quality of Lanesville TV reception from home to home. We had our loyal fans, but no one was about to adjust his or her reception exclusively for our broadcasts.

∞∞∞∞∞∞

Doyle told us our first broadcast reached as far away as friends of his in Kingston. Broadcast transmissions in the VHF band do have a way of playing funny tricks, skipping off the atmosphere and ending up far from where you intend them to be seen. I find it hard to believe we went beyond the confines of the Stony Clove Valley that night. And if our signal that night did escape to a wider world, we never repeated the feat. More often, people as close as Sam and Miriam reported difficulties with reception, and we usually devoted a portion of each show to fine tuning the persnickety transmission system, with much shouting up and down the stairwell between first and third floors as viewers told us whether Chuck's tweaking made their reception better or worse.

Our telephone connection to the community started as an adjunct to the shows, not a part of them. We'd call people we knew had tuned us in and ask them technical questions. We'd also solicit calls from the community. At first, one of us would handle those calls in the viewing room, with the information passed on to Chuck via lung-powered intercom or, if it had to do with the programming, to the people running the show in the control room. But very soon, we realized we should put the voices out over the air so that everyone could hear what their neighbors had to say. TV stations didn't broadcast phone calls live in those days because they feared some caller might say something that would jeopardize their government licenses. We had no license and nothing to lose.

Broadcasting the calls also served to reassure the community that Lanesville TV included them, that our neighbors had a voice in our TV station. Their voices, more than anything else, made Channel 3 different in our minds from any other station. Over the course of the five years we remained on the air, the phone became an emblem of the openness and responsiveness of the station to concerns of the viewers. And it worked well in the beginning. We got our share of crank calls and silly requests, but people also called to announce community events or to discuss the contents of the shows. In later years whenever no one phoned, we'd resort to pleading for calls just to let us know someone had tuned us in. Sam almost always responded, and if he didn't, one of us would call him.

And when calls did come in, we didn't always get the response we hoped for. One time Gert called to say she wanted to watch *All in the Family* and our Channel 3 interfered with her reception of Channel 3 from Hartford, which carried Archie Bunker. No one called to request we continue that night, so we pulled the plug for the evening and ate dinner early.

Another time, a ham radio operator called to say we were interfering with his reception of *Wild Kingdom*. "You're on every channel," he complained.

Nancy hosted the show that night and she kept her cool. This happened in April of 1976 after four years on the air. By that time, all of us had ample experience with hostile callers. First things first. "Do you consider yourself a viewer of us or an avoider?" she asked with a note of hope in her voice.

"An avoider," he said without hesitation. He wanted to know about our license. Visions of J.P. and the FCC gumshoes. This could lead to trouble. "You're an experimental station?" he asked.

Nancy pondered that a moment: "We're an experimental station," she nodded.

Yes and no. Never once that I recall did we identify ourselves in Lanesville as a pirate TV station. We were simply Lanesville TV and not about to pirate anything from anyone in the Stony Clove Valley. What made us pirates was that not only did we have no license to broadcast, we couldn't obtain one. Skip had written to the Federal Communications Commission about the time we went the air, discreetly inquiring about the possibility of applying for a low power TV station license. He was told no such license existed. Nearly a decade later, Lanesville TV became one model on which the FCC based its new Low Power TV Broadcast Service, but while Lanesville TV was on the air, it was unlicensed and therefore illegal in the eyes of the federal agency charged with regulating use of the airwaves.

The FCC did have a class of station known as experimental, but we wouldn't have qualified for that technically narrow category. One small loophole existed, one that not only fit us, but which had allowed the Catskills to become the site of the first legal low power TV station in the country long before we'd arrived. No one at the FCC volunteered to tell us anything about that station and we didn't learn of it on our own until we'd broadcast for several years and didn't care anymore about seeking a license.

So when it came to people like our neighbor, the ham, we maintained the fiction that we operated an "experimental station" in the general sense that we were experimenting with broadcasting while flouting the regulations meant to prevent just this type of experimentation. "You're not allowed to louse up the spectrum," he grumbled. Smiling sweetly, Nancy wrote down his advice on how to correct the problem. After the show, Chuck and I renewed our efforts to reduce the kinds of undesirable emissions that caused our signal to distort or altogether block out other stations–the way Abbie had originally intended for us to do.

Either Chuck was successful or the ham changed his viewing habits. He didn't call again to complain, and the Commission's Field Operations Bureau never broke down our door. The FCC has no record of any complaint about Lanesville TV. But I found calls like the one from the ham discouraging. I took it personally whenever we learned someone in Lanesville chose not to watch our broadcasts. I felt better, though, with his admission he was a weekender, not a full-time Lanesvillian.

∞∞∞∞∞∞

We didn't enjoy universal popularity, even among our year-round viewers. The wife of the owner of the largest business in town, the sawmill, advised some people we knew not to let us into their homes for interviews. She said she wouldn't let us into her house for fear we wanted to tape things we planned to steal. Had we cared about our reputation in the community, we might have worried about that kind of small-mindedness. But when Lanesville TV began, we had no kids in school, belonged to no local organizations and our interaction with town government had consisted of Carol and me getting married by Covey Chapman.

We'd had no interaction with local law enforcement, either, such as it was in Lanesville. The same could not be said of cops in surrounding communities. I'd had my first encounter with Shandaken constables the evening we signed the lease for Maple Tree Farm. Carol and I had gone to Woodstock for dinner and on our way back to Lanesville to spend our first night there, a cop pulled us over in Phoenicia. I was driving my beat-up 1963 VW bus, the quintessential hippie vehicle, which had attracted a number of unprovoked encounters with cops, several of them ugly. My bus had no particularly outstanding markings, no psychedelic swirls or anti-war slogans, just the name "Luis" scratched on the side in rather small letters by a timid graffiti artist, a souvenir from the days when I parked outside David's and Curtis' loft on Rivington Street.

The constable, a compact man with a sunburned face, sounded more businesslike than hostile. He asked politely for my license and registration, checked the tires and the blinkers, the kind of a welcome-to-the-neighborhood routine I'd become accustomed to. When cops couldn't find something easy to nail hippies for, they often turned surly and made up a reason to hassle you. A bored state trooper had hauled me into his barracks outside Albany a year earlier and strip-searched me for no apparent reason and with no result. And right after Woodstock a bull of a New Jersey state cop pulled me over on the turnpike and extorted $75 when he couldn't find anything valid to bust me for. I was ready for the worst in Phoenicia, but this time it was different. The cop handed me my documents and walked away with a nod that looked almost friendly.

The next time we met, I had him in my camera viewfinder a few weeks after Lanesville TV went on the air. I had gone out to tape the crew from the Stony Clove Rod and Gun Club stocking the stream in advance of the first day of trout season. The club was the biggest organization and the only one of any consequence in the valley. Its fall turkey shoot and summer clam bake were the social events of the year in these parts, and many local landowners had agreed to allow the club to post their land, restricting hunting and fishing to members only. Holding office in the club gave a man far more stature in these parts than becoming a town official. In many respects the club governed the valley. The treasurer of the club and its spokesman when we arrived was a native of Scotland named Hugh Stuart, though no one ever called him anything but Scotty. He made a living as a house painter and wallpaperer, and he spoke with a thick burr, which made him about as exotic as anyone could hope to be in Lanesville and still win acceptance. Scotty and his family had become regular viewers of Lanesville TV. They had good reception because they lived only a few houses down from us in one of the only dwellings with an unobstructed shot at our antenna. Scotty had alerted us to the upcoming trout stocking ritual, and when I showed up to tape it, he introduced me to the cop, Shandaken Chief Constable Jack Schlegel. Jack was in charge of the stocking operation.

The trout arrived from the hatchery in iron tanks on the back of a pick-up. Even if they're born in captivity, the types of trout that give the streams of the Catskills their international reputation for sport fishing need lots of oxygen in order to survive, so the pick-up also carried a pump to aerate the tanks. The pump motor and the burbling water it produced created a din around the truck that heightened the excitement of the occasion by forcing everyone to shout at each other to be heard. As the truck drove up Route 214 to predetermined stops, club members grabbed plastic buckets full of froth and writhing trout, looking like so much animated jewelry. The men trudged through patches of snow that still covered much of the ground in Lanesville in late March and dumped the fish unceremoniously into the stream. As the tedious process unfolded Jack got to talking.

I've never met anyone more at ease in front of a camera. Fully aware the tape was rolling, he flourished the mic as if he'd been born with it in his hand, expounding passionately on the importance of the club's work. As part of this peroration he added some uncomplimentary observations on the lowland fishermen who cast their lines in Phoenicia and reaped a windfall of club-stocked fish which had migrated downstream. There was something inherently unfair in that, according to Jack.

The stocking tape played a few nights later on Lanesville TV, and several club members called in, amazed at having seen themselves on their own TVs. Others stopped me at the general store to say what a good thing it was to have the club on TV. About time, too. It was our first major breakthrough in the community and the only show many people in Lanesville would ever admit having seen.

Others took a less sanguine view of that segment. The Phoenicia fishing fraternity roundly criticized Jack for badmouthing downstream anglers and for disloyalty to the town that employed him. He had cast his lot with that clannish group in Lanesville, a place that wasn't even a part of Ulster County. As he was helping clean up after a local fire four years later, Jack declined to be interviewed, good naturedly recalling how those trout stocking statements "got me in trouble." Neither he nor anyone else questioned how he had come to be on TV. What they cared about was what he'd said.

7
TVTV

Since we'd decided to stay together as a group after the end of our show for Don West and CBS, our individual styles of living and working had evolved informally. We Upper West Side apartment people, Carol, Nancy, Skip, Bart and I, converged with the Lower East Side loft dwellers, David and Curtis, at our common turf, the Soho studio, trying to adjust our individual and collective projects around the pleas of Chuck, Ann and Davidson for a little privacy. All around the studio, meanwhile, the neighborhood was undergoing rapid change. The baking company a block away that had filled Prince Street each night with the succulent smell of hot, spongy white bread shut its doors for good one day, just another casualty on a long list of factories and warehouses that had once lined these narrow streets. Into the void rushed gallery owners like Reese Paley, known for his marketing of ceramic birds and later, on Prince Street, for his champagne fountain openings and blood sculpture shows. It was no longer unusual to see people walking the streets on weekends, women in furs, gray-haired men in expensive topcoats. They parked Volvos and Cadillacs with two tires on the sidewalk. Soho had been discovered. Only a matter of time before the same happened to us, too.

We made a few tentative attempts to discipline ourselves in preparation, deciding in the summer of 1970 we needed weekly meetings of the group; attendance required. We gathered in a circle at the loft and talked and talked and talked. Within the next few days I found a job that called me away at meeting times. Other Freex scowled: There'd better be money involved. There was, a paying gig as cameraman and technical consultant for a program documenting Council projects around the city, taping the Alvin Ailey Dance Company, riding with the Jazzmobile through Harlem. Meanwhile, others took my lead, and after a few weeks, it seemed nobody was free for meetings.

The move to Lanesville carried with it the promise of a new discipline imposed by the very fact of living together in such close quarters. We agreed the first week at Maple Tree Farm to discuss all important matters at dinner meetings of the full group, hoping this would ease the pain of the process and that appetites would enforce compulsory attendance. We reached two memorable decisions at those meals. The first was to cancel David's invitation to Bob Quinn and his family to join Videofreex. The second was to name the dog.

I had dubbed him Meatball when Carol and Nancy brought him home to the apartment from the vet. Once we got to Lanesville, some Freex felt Meatball was not a dignified enough name. They threw out suggestions. "Rushmore," said Frances.

"Did you say MUSH-room?" asked Davidson, squinting in disbelief.

"Mushroom!" It was a general cry around the table. Davidson looked skeptical, but he was overruled. Meatball became Mushroom by acclamation.

So much for the benefits of collective meetings. Rehashing the minutiae of group life held little appeal when we could be editing or shooting tape, plotting privately or playing outdoors. The tasks required to keep the farm running devolved to the people willing to take responsibility. Bickering over the identity of the culprit who left dirty dishes in the sink or hogged the last blank tape in the second kitchen did not require group adjudication, and substantive questions of direction or focus never waited for formal review.

∞∞∞∞∞

Sarah was the first of the Freex to claim membership by birth, not that she cared. She arrived on a muggy morning in late May, almost a year after we moved to Lanesville. Unlike the adventurous folks at Rainbow Farm in Phoenicia, Carol had no fantasies about delivering at home. She went into labor, woke me up and off we drove at dawn to the hospital. The arrival of seven-pound-five-ounce Sarah Nell Kellough Teasdale was dutifully announced by a special broadcast on Lanesville TV.

Everyone had looked forward to Sarah's arrival, and everybody but Chuck, who reacted to children as if they were faultily designed appliances, helped take care of her from time to time. Other than some food in the fridge declared off limits to adults, having an infant in the house didn't outwardly change the rhythms of life at Maple Tree Farm. But my life and Carol's had taken on a new focus, one we couldn't share equally with the rest of the Videofreex.

Sarah's birth found me in the midst of writing my book about portapaks. That kept me close to home with ample time for fatherhood. I got up with Sarah in the morning when she was old enough to eat solid food and fed her breakfast in a silent kitchen. Everyone else slept late, just as I used to.

I had afternoons to myself. While other Freex shot tapes, worked in the editing room or plotted new projects, I trudged up to the third floor and shut the door of my cubbyhole office, where I scribbled away with unreliable ballpoint pens from Phoenix House, a big drug rehab program in Manhattan. Phoenix House didn't have much cash, but it had a warehouse of donated goods, which its leaders opened up to us as payment for workshops the Freex had done for the program. We ended up with huge cans of government surplus peanut butter and 50-pound bags of unroasted coffee beans. The food disappeared quickly what with all our guests. The pens, unfortunately, lingered.

During the late winter and early spring of 1973, I'd have to stop writing every so often to massage the cramps in my hand. During those pauses I'd watch the spectacle in the meadow next door. The players were Gene and his employer, Walt Bauman, a rough looking, heavyset man, who had growled something at me once. I never had any other communication with him. He lived in a trailer at the far end of the meadow next to the Farm. Sarah and I watched him in the mornings when he drove his Caterpillar bulldozer up the mountain, with his chain saws and a case of beer strapped on the back and Gene slogging along behind. In the late afternoons from my third floor lookout, I saw the two men emerge from the forest with the day's cut. I'd been told a story, probably apocryphal, that Walt shot a deer one night out the back window of his trailer. He dragged the animal into the bathroom so no one would see he'd jacked a doe, and done it out of season, too. He planned to dress it after dinner, but while he was eating, the animal revived and began to raise hell in the bathroom. He couldn't shoot it at such close quarters inside his own home, so Walt had to finish it off with his hunting knife, which he did, but only after a gory struggle that leaves me imagining the shower scene in Alfred Hitchcock's *Psycho*.

So here came Walt, burly, taciturn and a good man with a knife, maneuvering that bulldozer through the mud to make neat piles of the thirty-foot oak trunks. He nudged each huge tree precisely into place as if it were a child's toy, and as I watched, I began to see his dexterity on that machine as a dance, intricately, even delicately choreographed, and efficient in the extreme. One day he didn't show up and the logs lay by the side of the road for months. I heard a tree had fallen the wrong way and struck him on the head, injuring him badly. In an offhand way, I laughed with people who said his head was harder than the log. Back in my office, I missed the performances.

I had no idea how long it took to write a book. The research portion had come naturally. I'd hung over Chuck's shoulder for so long, I'd absorbed the rudiments of video maintenance. Some of it looked like easy stuff that shouldn't require an expensive trip to the technician. But how would anyone know that? Before we left the city, I wrote a piece for the Raindance publication, *Radical Software*, on how to change the portapak camera tube. I saw that procedure as the equivalent of changing a car tire, something everyone who drives should know how to do. Perpetuating the mystique that made minor repairs like changing tubes look impossibly complex caused people to spend money on maintenance they might better have spent on production. I decided to expand that magazine piece into a book on how portapak technology worked and how untrained people could keep it working. I modeled the book on a cleverly illustrated paperback about car repair, called *Volkswagens for Idiots*, which was written in a low key, non-technical manner. I had relied on that volume to keep my bus, Luis, on the road.

∞∞∞∞∞

My decision to concentrate on finishing the book helped me adjust to the new restrictions parenthood placed on my freedom, though it didn't take long for me to feel the bite of those self-imposed restraints. Little more than a month after Sarah arrived, Nancy, Bart, Skip and Chuck began preparing to leave for Miami to work with an ad hoc group called Top Value Television, TVTV for short, covering the Democratic National Convention in Miami. TVTV planned to put video freaks from all over the country in the midst of the national political process and see what came of it. In that sense, this first TVTV project borrowed on the experience of the May Day Collective and even *Subject to Change*. But unlike those two productions–or perhaps because of the lessons they'd taught us all–TVTV had done considerable advance planning.

Michael Shamberg gets the credit for coming up with the idea for TVTV. He had written for *Time* and *Life* magazines, and he and several other people, including Ira Schneider (who'd informed us of Bart's Videofreex ties before Bart had), Megan Williams, Frank Gillette and Beryl Korot had started Raindance in 1970. As a group, they were as interested in analyzing the new media phenomenon of video as they were in making tapes. Out of this division of interests came the publication *Radical Software,* which quickly claimed the vacant title of literary and intellectual voice of the video movement. One of the chief editors of *Radical Software* was Phyllis Gershuny, a tall woman with a deep voice, who moved into Prince Street with Davidson, sharing the black vinyl tent he'd draped from the sprinkler pipes in the shooting set area of the studio. Phyllis arrived pregnant and left a few months later after giving birth to a daughter. I felt a sense of relief at the departure of mother and infant; better for the child not to live like that, I thought, and better for us not to have someone so closely tied to another video group privy to all our messy Freex struggles.

Radical Software, a thick tabloid, made good use of the Raindance group's sense of graphic design. Most other video freaks had barely mastered the electric typewriter, and the large illustrations and crisp typefaces of Raindance printed materials looked slick in comparison to efforts like our short-lived *Maple Tree Farm Report* newsletter. And the artistic success of the publication fed our suspicions that Michael, in particular, was trying to straddle the incompatible worlds of freak media and straight. These feelings were only heightened when he suddenly went to Vietnam for a few weeks with a portapak. None of us had his type of press contacts, and there was something disquieting about his access to military credentials, but most of all his trip to the war zone seemed like a wasted opportunity. He'd returned with nothing more to show than a few brief interviews with a handful of shirtless soldiers at a well entrenched artillery battery, who fired occasional rounds somewhere into the jungle. The episode confirmed my belief Michael didn't have much of a flare for shooting tapes, but that was never a skill he pretended to have.

He did know how to write well, and his book, *Guerrilla Television,* which came out in late 1971, was the first book written about the video scene from an insider's point of view. It drew national attention to the video movement and legitimized all us video freaks in the eyes of the conventional press to a degree we could not have hoped to achieve with our tapes alone. But many of us "guerrillas" mistrusted Michael's motives. His attenuated vowels, penny loafers and Polo shirts stood out to us as clues that preppie values superseded his commitment to a revolutionary change in the media. We had difficulty appreciating *Guerrilla Television* as the most articulate statement to date of what all of us were doing with video. And it rankled that he could write a book dismissing such important players in the video movement as Videofreex with thumbnail sketches. We thought we deserved more attention.

For the conventions, Michael not only had a good idea, he had a few thousand dollars in grants, just enough money to start putting the plan in place. The venture was to be co-produced by Raindance and the San Francisco group Ant Farm. From the beginning Michael was the most visible member and frequently the only one quoted in the press. He and Tom Weinberg, a video producer from Chicago, and a few others working on the project came up with airline tickets, a place to stay in Miami and lots of blank tape. Most important, they managed to score legitimate press credentials, which would allow TVTV crews to roam the floor of the convention with their portapaks. That was unprecedented; until 1972 TV footage of the convention floor was the exclusive province of the three networks. The TVTV plan envisioned separate teams following different aspects of the event: the politics, the media, the demonstrators, etc., with each crew responsible for its own equipment. Michael had no guarantee of a market for the tapes, only an expression of interest from executives at several different cable TV systems. The executives said they wanted to see an edited version of what TVTV shot in Miami.

Carol and I talked it over. She couldn't go because she was nursing Sarah. I wouldn't go because she and Sarah needed me at home, and we convinced ourselves I didn't want to work for Michael, anyway. I bowed out of the production phase, while agreeing to help with the edit.

The Freex who'd agreed to go to Miami departed toward the end of July. The month before, police arrested five men who had broken into the offices of the Democratic National Committee in Washington, D.C., at a building called The Watergate. Nobody paid much attention. We assumed Nixon's agents had a hand in it. They–the government–had taken photos of our license plates and tapped our phone, hadn't they? Nothing new about a break-in. Keep the focus on the war, we told ourselves. Nixon had announced the withdrawal of U.S. ground combat units from Vietnam. At the same time he increased the bombing of North and South Vietnam to horrific proportions without noticeable effect on the advance of the communists. The Democrats would nominate South Dakota Senator George McGovern for president, a colorless man who promised to end the war. What a change from four years earlier, when Vice President Hubert Humphrey stood idly by while Chicago police rioted, beating and gassing antiwar protesters in front of the Democratic Convention. Our tapes of the Chicago Seven during their trial for conspiring to organize the '68 demonstrations lay at the core of the Videofreex mythology. Somebody had to represent us in Miami.

This time no police would riot. The demonstrators became part of the convention. Young political activists with strong anti-war convictions had forced changes in party rules and opened up control of the nominating process. In July of '72, though, the implications of the changes in the Democratic Party were not yet clear, and no one, especially the national press corps, could predict what would happen at the convention. Chicago had produced a great news story, particularly for CBS, which broke away from the convention floor to show tapes of middle-class white kids being clubbed by the cops. Others had turned a blind eye as long as they could. Now the network reporters descended on Miami, ready to report it all without understanding they no longer possessed a monopoly. They strutted around as vain and pompous as any politician, totally unprepared for TVTV, which uncovered how they covered the convention for America.

As soon as the convention was over, the TVTV production crews disbanded. Nancy, Bart, Chuck and Skip returned to the Farm with copies of some of their tapes and plenty of stories to tell from the convention floor and the high life in Miami. I felt a twinge of envy. We played some of the tapes, including long interviews with media celebrities like NBC news anchor John Chancellor, on Lanesville TV. Meanwhile, Michael and other TVTV organizers watched the tapes and hastily planned an edited version they called *The World's Largest TV Studio*, an hour-long documentary with no narrator. It captured a convention in a way never before attempted, intimately and with attention paid to humanity and political mechanics rather than to the grand scale of the spectacle.

Among the scenes the TVTV crew had access to were caucuses of some of the key state delegations whose leaders were planning and executing sophisticated political strategies. The strategies were so complex in some cases, that following what California state Legislator Willie Brown (now mayor of San Francisco), explains to his supporters would tax even a political scientist. What no viewer would miss in this tape and many other TVTV sequences is the group's startling level of access.

For logistical, and perhaps political reasons of his own, Michael chose not to edit the tapes at Maple Tree Farm. Instead, he picked a small production studio still under construction in lower Manhattan called the Egg Store. It was owned by a video equipment dealer named Chi Tien Lui, known among video freaks as "Cheatin' Louie." I doubt Lui behaved any more disreputably as a businessman than his competitors–one of them, Sammy Adwar, was caught by a *60 Minutes* undercover camera a few years ago hawking bootleg movies on tape–but Lui was definitely the most colorful character among the small, highly competitive fraternity of Manhattan video equipment dealers. According to Chuck, Lui jumped ship from a Taiwanese freighter, a story he told Chuck when the two of them worked beside each other as bench technicians repairing video equipment. Lui went into business for himself and had some success selling to the burgeoning video movement, in part because he acted hipper than the other dealers, was known to attend the right parties and wore bright colored suits with broad bell-bottoms. My most cherished image of this wily video entrepreneur is on tape: him standing alone on the balcony in the loft of a Manhattan designer of movie titles, Pablo Ferro, during a raucous party on New Year's Eve 1969, an event known at the time among video buffs as "the orgy." Lui looks down at the camera and smiles shyly, wearing nothing but his socks.

The Egg Store was a dark cavern a few short blocks from the Hudson. Its walls and ceiling had been sprayed with clumps of a gray, fibrous material that only intensified the gloominess of the space. A technician who worked for Lui claimed years later that the fibers were asbestos. The control room, where the editing machines were set up was so narrow only a handful of people could work comfortably at one time. So we set up monitors in the studio, where TVTV members not involved in editing the current sequence in the control room could view other raw footage and argue over what should come next.

I practically lived at the Egg Store for more than a week as the technical advisor to the edit. My skills as a techie shone only in comparison to the majority of video freaks, who either maintained a willful ignorance about the workings of the medium or pleaded congenital ineptitude. My real skill lay in my ability to talk to technicians without berating or snubbing them, which meant I could usually get straight answers I then interpreted for others. I encouraged Michael to put his faith in technical wizardry I did not possess because it gave me a unique role. I could stand aside from the bickering over which sequence should go where, only to offer, when requested, a considered opinion based solely, you understand, on technical considerations.

Tom, Ira, Allen Rucker and other TVTV participants who came and went during the editing of the Democratic Convention tape, grew increasingly impatient with Michael's decisions. Some of them saw him as a budding autocrat, taking over a process they intended to be cooperative. They had no alternative ground rules for collective decision making, and many of them turned to me at one time or another to ask how Videofreex did it. I didn't have a ready answer, although I shared their dismay at Michael's control of the process. He was already thinking about plans for TVTV's coverage of the Republican National Convention, also scheduled for Miami, and had little patience for criticism of his methods. At one point I grumbled about his decisions while the two of us wrestled with a troublesome sequence. "Well, Parry, how would you have done it?" he demanded, funnelling his exasperation into a challenge.

"I wouldn't have shot all this tape," I said without hesitation. "I would have done it live."

What a lame response! Almost nothing but Lanesville TV existed as an uncensored outlet for TVTV's tapes, and those tapes deserved a far larger audience than we could provide. Cable TV public access channels amounted to little more than a promise. Broadcast executives wouldn't have returned Michael's calls had he sought to convince them to play TVTV's video as something other than a tightly packaged hour carefully scrutinized by censors and advertisers. Yet the necessity for packaging the event created a conundrum of its own: The value of the conventions appeared to lie in their appeal as news; packaging them meant the information they contained was outdated by the time the edits were done.

Four years later, during the 1976 Democratic Convention in New York, Tom, Videofreex, independent producer DeeDee Halleck and others formed a new ad hoc group called the Image Union, recreating the key features of Lanesville TV on a public access channel in Manhattan for a collectively produced series called *The Five Day Bicycle Race.* For five nights during prime time, we combined live segments with a phone on the air, as well as tapes shot by crews shuttling back and forth between the convention floor at Madison Square Garden and our tenement-like studio on East 23rd Street. But in 1972, none of that was possible from Miami. My answer to Michael reflected a political rigidity about the righteousness of Lanesville TV, not a practical solution to a brand new problem.

Michael and I had a clash of visions–his for taking the best of what we did and adapting it for a mass medium, and mine, which regarded Lanesville TV's unpackaged, community-responsive, seat-of-the-pants production approach as inherently superior and incompatible with the media establishment. I couldn't come up with anything better to tell him, either. He faced a real bind and I think he genuinely sought my advice. But he had appropriated for himself the role of producer. In TV terms, that made him the boss. My sympathies lay with the workers. He would have to get out of this one by himself.

He did, with bullying, compromise and sheer determination. We completed the edit of *The World's Largest TV Studio* on time, and TVTV followed it with *Four More Years*, a tape of the Republican Convention, which anointed Richard Nixon. The Nixon convention provided much starker and more marketable footage, with the party faithful marching in lockstep to the commands of their soon-to-be-disgraced leaders. The Nixon love-fest in Miami lent itself to images filtered through the wide angle-lens I had picked up from Grayson Mattingly with my first video equipment. The wide angle distorts slightly depending on the distance from the subject in view. It gives a fish-eye view of the world, elongating the nose and mouth of interview subjects and foreshortening the tops of their heads. We had used it frequently in our tapes and it became the visual signature of these and subsequent TVTV programs. At the GOP convention an Ant Farmer with a wide angle lens taped while he walked up an aisle of wildly cheering delegates until he reached the back of the hall, where crippled Vietnam veteran Ron Kovic shouted protests against the war. Oliver Stone recreated that sequence almost exactly in his film *Born on the Fourth of July*.

When the TVTV convention tapes played on cable TV in Manhattan (the rest of the city did not yet have cable) and on a few other large systems around the country, they won critical acclaim in the straight press. *New York Times* critic John J. O'Connor devoted a long column to *The World's Largest TV Studio*, concluding that it was "distinctive and valuable." Richard Reeves, writing in *New York* magazine, called it "the best electronic coverage of the Democratic Convention." Even *Women's Wear Daily* felt obliged to weigh in, comparing the Nixon convention to the 1934 Nazi Party rally in Nuremburg and describing TVTV's exposure of it in *Four More Years* as a "superbly photographed and edited documentary."

Raindance withered, and TVTV, no longer ad hoc, moved its base of operations to the West Coast and became a production company. It created new programs for the rest of the decade, with Michael Shamberg more and more in control and members of Videofreex, primarily Skip, Nancy and Bart, making significant contributions. In 1974, the Public Broadcasting System presented the TVTV production of *Lord of the Universe*, a video portrait of a 15-year-old Indian guru and his enraptured followers. It became the first program shot on half inch tape to be shown nationwide. Mike Dann of CBS had admonished us after *Subject to Change* in 1969 we wouldn't be ready for a network show for five years. *Lord of the Universe* fulfilled his prediction to the year.

By the time TVTV disbanded, it had produced a remarkable number of documentaries, including *Adland*, an inside look at the TV advertising business, a series called *Gerald Ford's America* about government in Washington in the mid-1970s, and a look at the hoopla surrounding the Super Bowl. Each tape became more polished, more knowing and less spontaneous. Most reached far larger audiences than the convention tapes. Eventually, Shamberg left video and went his own way, becoming a successful movie producer, responsible for Hollywood box-office hits like *The Big Chill* and *A Fish Called Wanda*.

VIDEOFREEX

A view through the summer sun porch at Maple Tree Farm, looking down Route 214 toward metropolitan Lanesville. The stained glass "Freex" piece hanging at left was created by "Horrible" Howard Raab. *(Media Bus archives)*

Looking across Route 214 at Maple Tree Farm the winter after Videofreex first arrived in Lanesville. *(Photograph by Davidson Gigliotti, from the author's collection)*

The author on the Lanesville TV set for the live, call-in segment of "The Bunny Returns" show in 1976. *(Media Bus archives)*

The Bunny informs Lanesville Player and resident Bobby Benjamin of the Easter egg shortage. *(Off-screen photo by the author)*

VIDEOFREEX

Two views of the Blue Calzone, our inflatable TV set, unveiled at the Alternative Media Conference at Goddard College in Vermont. The video projector was housed inside, with the image presented to people outside on the rear projection screen mounted at the front. It also served as a tent for some Freex during the conference. *(Media Bus archives)*

VIDEOFREEX

In one of the early broadcasts of Lanesville TV, Elmer Benjamin ponders Carol's request that he name his 17 children. He needed some prompting. *(Media Bus archives)*

In the initial episode of Bart's cowboy show, the three regulars, (from left) Buckaroo Bart (Friedman), Horrible Howard (Rabb) and Sheriff John (Benjamin), stroll down Neal Road just before Horrible's shocking criminal act of littering. *(Off-screen photo by the author)*

93

VIDEOFREEX

Davidson as Sam Trouda, the uncatchable trout, taunting local anglers on Lanesville TV. *(Off-screen photo by the author)*

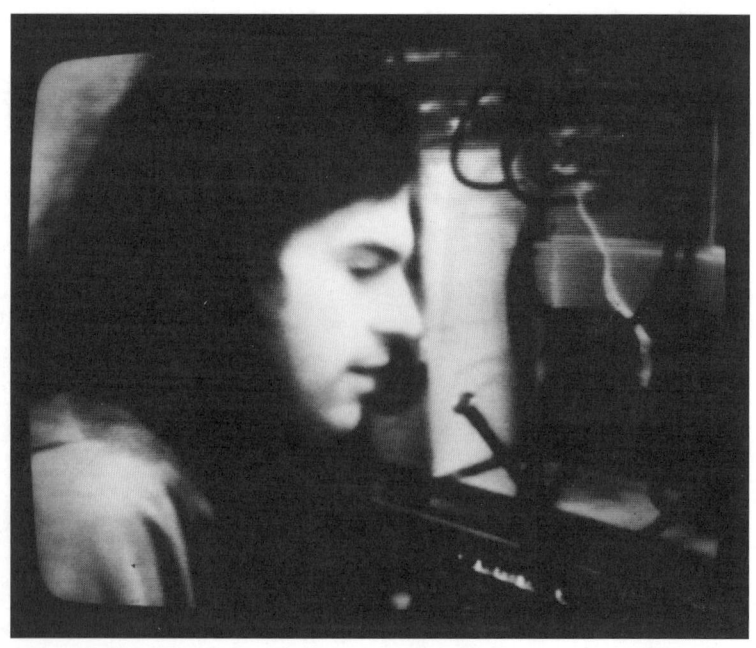

The author hosting the first Lanesville TV pirate broadcast, March 19, 1972, probably just moments before the old gooseneck mic, liberated from CBS, became entangled in his hair. *(Off-screen photo by the author)*

VIDEOFREEX

On the set of Lanesville TV for our first show broadcast not only in Lanesville but also, later, on public television in New York City and elsewhere around the country on the "Video and Television Review" series. Nancy Cain takes a call from local patriarch Willie Benjamin, while Harriet Benjamin, wearing a wig, reacts and Russell Connor, who headed the video program for the New York State Council on the Arts and co-hosted this show, looks on. *(Media Bus archives)*

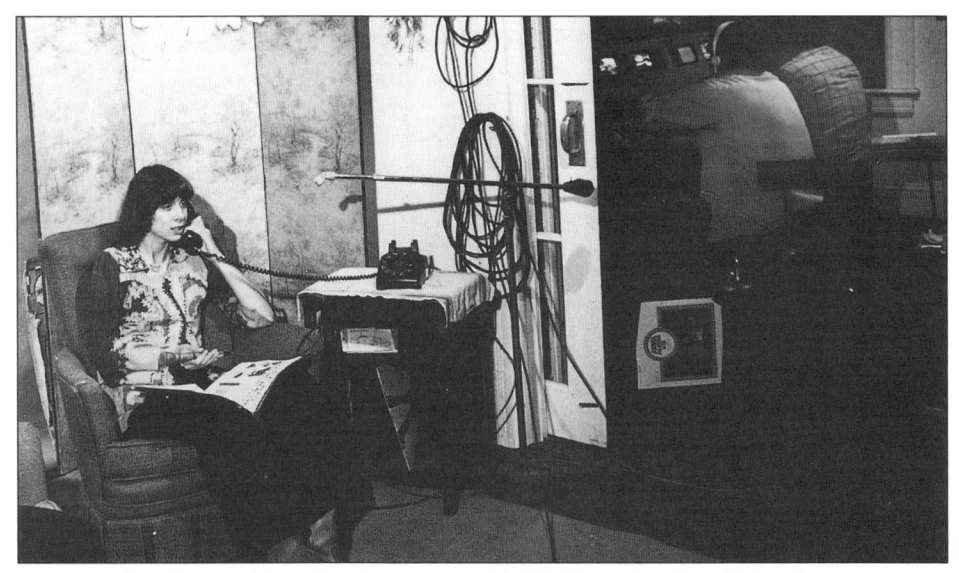

Nancy Cain takes a call on the air at Lanesville TV. Note that the sliding doors between the set and the control room are open—the usual style for the show and the easiest way to get around the need for an intercom. In the control room, with their backs to the camera, are Chuck and Bart. *(Media Bus archives)*

VIDEOFREEX

The heart of America's smallest (and most illegal) TV station, Lanesville TV, Channel 3, was the control room. Here, with their backs to the camera, Chuck (l) and Bart engineer a broadcast, which required at least two people but no more than three. Davidson built the plywood console, driving off the neighbors in the process. The low stand at Bart's right holds an old hotel reservations keyboard that Chuck rigged to let us insert text on the screen. *(Media Bus archives)*

The three monitors of "Hunter Mountain," one of Davidson's early video sculptures created at Maple Tree Farm. *(Davidson Gigliotti, Media Bus archives)*

The Lanesville TV transmitter, which drew its power from a wall socket and which, with all its pieces, was about half the size of a toaster oven. The total cost of the transmitter, minus the Sony battery charger (the second box from the bottom), which supplied the voltage for the final stage, was about $100. *(Photo by the author)*

Joseph Paul's original, homemade Lanesville TV broadcast antenna is visible in the background, center, between two commercial UHF TV receiving antennas set on a lower roof at Maple Tree Farm so we could watch TV when we weren't making it. *(Photo by the author)*

VIDEOFREEX

Bart in one of the mirror-image special effects used in "Oriental Magic Show." If you cover one side of his face, you get an idea how the image was created. *(Media Bus archives)*

David with Mushroom after one of Mushroom's confrontations with cable-munching porcupines. *(Media Bus archives)*

Close-up of Pedro Lujan's rope bridge over the Stony Clove Creek. *(Media Bus archives)*

Frank "the Fist" Farkle (right) and "Rocky" Van battle it out in Tannersville, with Videofreex taping every moment of the charity bout. *(Media Bus archives)*

VIDEOFREEX

Davidson, not in costume, sets up for one of his multi-channel video works: nude dancers. *(Media Bus archives)*

David in his ascetic phase, with Oberon perched in the branch at right. *(Photograph by John Dominis)*

The Lanesville TV logo designed by Nancy's father, Leon Wayburn. *(Media Bus archives)*

VIDEOFREEX

Academy Award winning director Shirley Clarke produces her video version of a Marx Brothers movie right in the middle of preparations for the wedding of Curtis Ratcliff and Cy Griffin at Maple Tree Farm. She persuaded Freex, from right, Skip Blumberg (Harpo), Alan Sholom (Groucho) and Bart Friedman (Chico) to improvise according to a script only she could envision, eventually wearing out the cast and, nearly, her welcome at the farm. *(Photo by Larry Gale, Media Bus archives)*

Just before the wedding, Skip pops out a third floor window, responding to directions shouted by Shirley Clarke. Hung over and worse, the author sags against the side of Maple Tree Farm on a porch roof a story below (center left in white shirt), no doubt wrestling with mental preparations for his first and last appearance as a man of the cloth. *(Media Bus archives)*

Davidson's sign announcing Maple Tree Farm. At right, Route 214 snakes up the Stony Clove valley. *(Media Bus archives)*

Maple Tree Farm in 1971, the summer Videofreex arrived. *(Media Bus archives)*

The Videofreex in Annie's garden in the summer of 1972. From left, Ann Woodward (holding Chuck's cat, Leo), Chuck Kennedy, Davidson Gigliotti, David Cort, Nancy Cain, Carol Vontobel, Bart Friedman, the author and Skip Blumberg. *(Media Bus archives)*

Bart, Nancy and Chuck pose for a farewell photo with Maple Tree Farm landlord Sam Ginsberg on the day the Videofreex finally left Lanesville. The last of the Channel 3 antennas, long unused, sticks out of the back of the pickup behind Sam. *(Photo by the author)*

VIDEOFREEX

Harriet, Todd and Willie Benjamin. *(Media Bus archives)*

Videofreex Gothic, a June 1972 photo of Carol Vontobel, the author and Sarah Nell Kellough Teasdale, the first new member of the group in some time. *(Collection of the author)*

The author in 1971. *(Collection of the author)*

The author and "Rabbi" Joe Keley at Willie Benjamin's birthday party in Lanesville. *(Collection of the author)*

The Videofreex circa 1970 in our Prince Street, Manhattan, loft. Standing, from left, Skip Blumberg, Chuck Kennedy, Davidson Gigliotti, the author and David Cort. Seated, Bart Friedman, Carol Vontobel, Nancy Cain and Ann Woodward. *(Photograph by Bill Cox)*

Chuck at his workbench, inventing new equipment and keeping the old stuff running between Lucky Strikes.*(Photograph by John Dominis)*

VIDEOFREEX

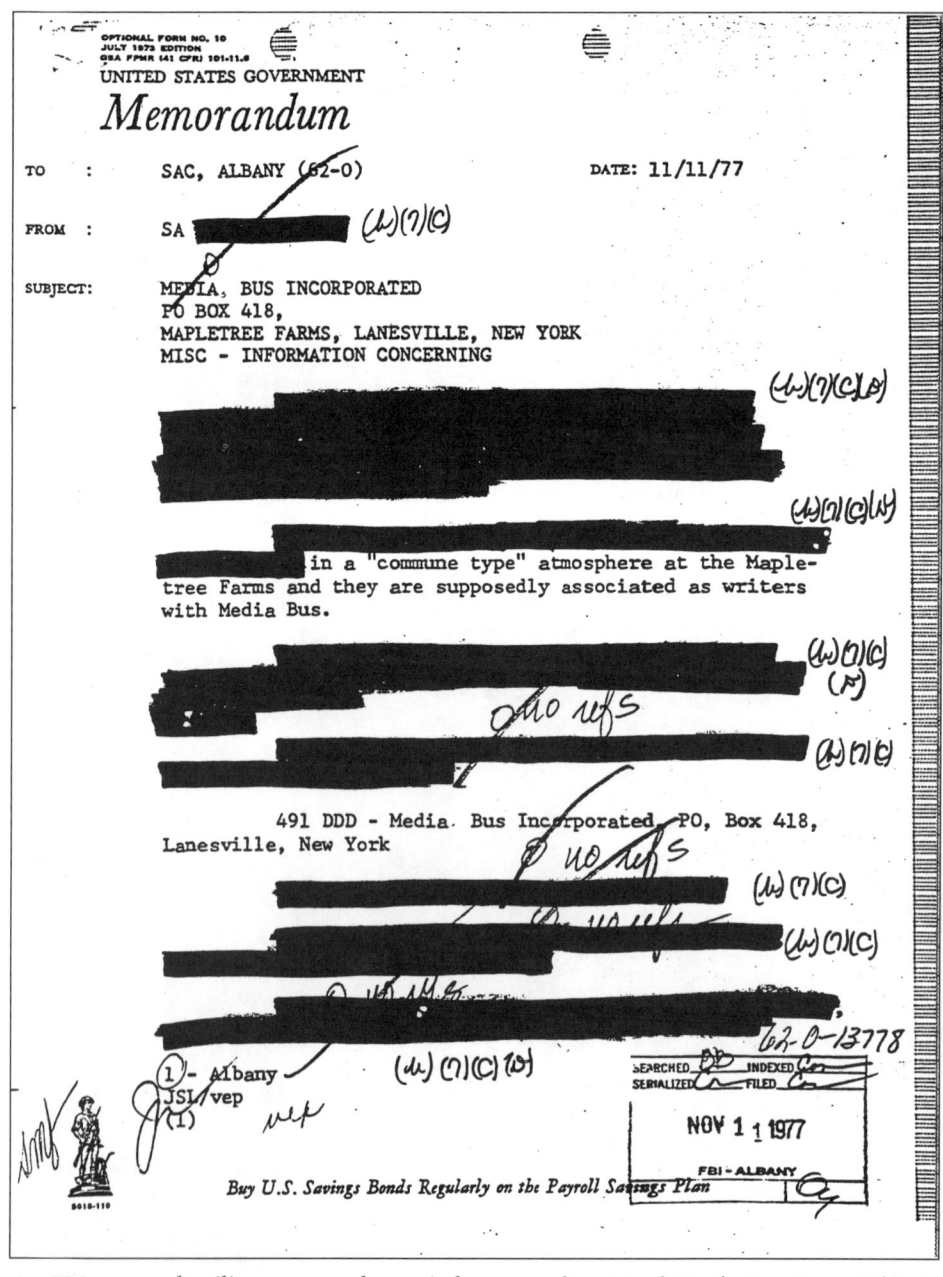

```
OPTIONAL FORM NO. 10
JULY 1973 EDITION
GSA FPMR (41 CFR) 101-11.6
UNITED STATES GOVERNMENT
```
Memorandum

TO : SAC, ALBANY (62-0) DATE: 11/11/77

FROM : SA ▓▓▓▓▓▓ (u)(7)(C)

SUBJECT: MEDIA, BUS INCORPORATED
PO BOX 418,
MAPLETREE FARMS, LANESVILLE, NEW YORK
MISC - INFORMATION CONCERNING

[redacted] in a "commune type" atmosphere at the Mapletree Farms and they are supposedly associated as writers with Media Bus.

491 DDD - Media Bus Incorporated, PO, Box 418, Lanesville, New York

62-0-13778

NOV 11 1977
FBI - ALBANY

1 - Albany
JSI/vep
(1)

Buy U.S. Savings Bonds Regularly on the Payroll Savings Plan

An FBI memo detailing a report by an informant who visited Maple Tree Farm. This is how the FBI originally released the document. *(Collection of the author)*

VIDEOFREEX

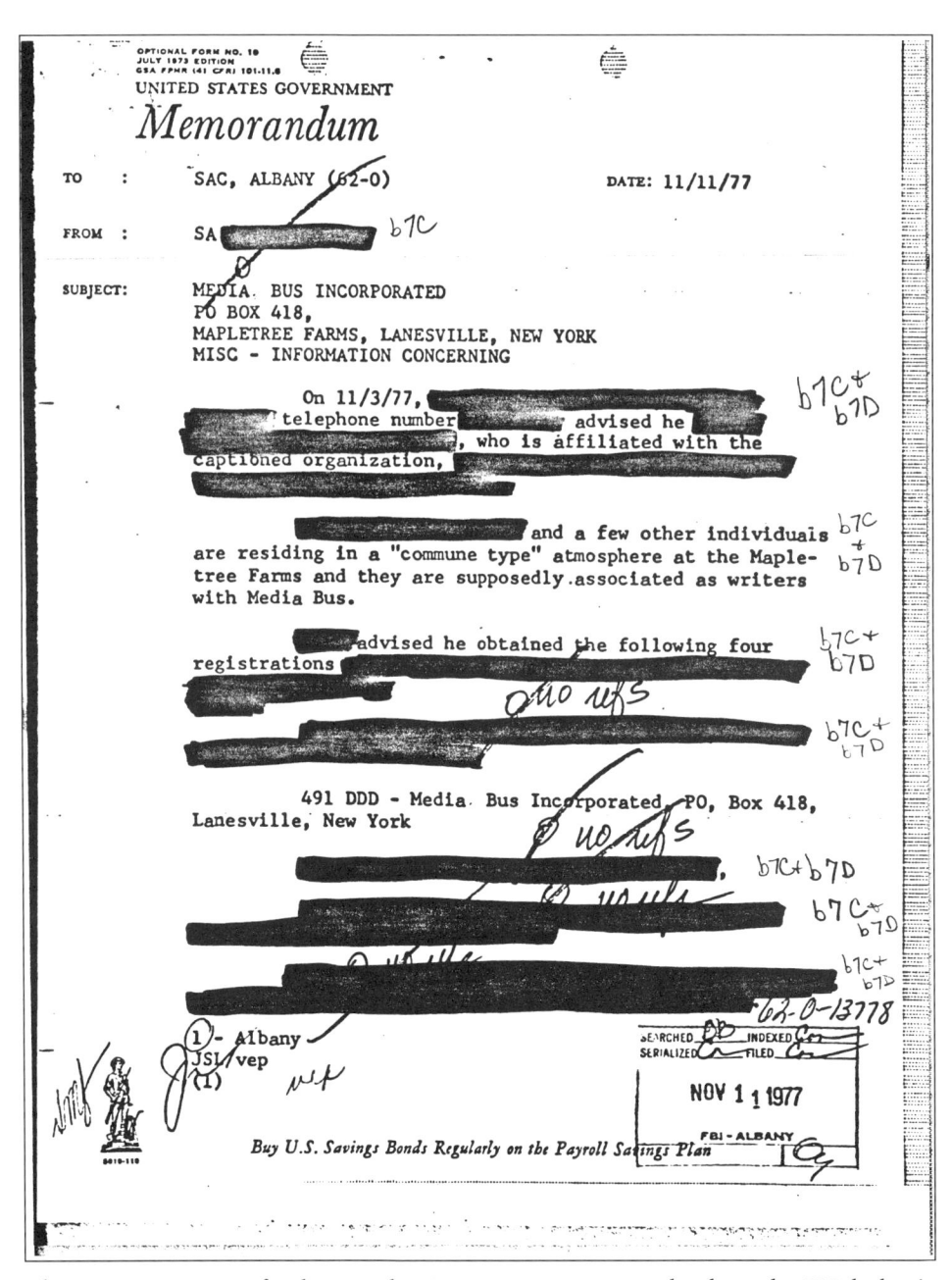

```
OPTIONAL FORM NO. 10
JULY 1973 EDITION
GSA FPMR (41 CFR) 101-11.6
UNITED STATES GOVERNMENT
Memorandum

TO     :  SAC, ALBANY (62-0)              DATE: 11/11/77

FROM   :  SA [redacted]  b7C

SUBJECT:  MEDIA BUS INCORPORATED
          PO BOX 418,
          MAPLETREE FARMS, LANESVILLE, NEW YORK
          MISC - INFORMATION CONCERNING
```

On 11/3/77, [redacted] telephone number [redacted] advised he [redacted], who is affiliated with the captioned organization, [b7C+b7D]

[redacted] and a few other individuals are residing in a "commune type" atmosphere at the Mapletree Farms and they are supposedly associated as writers with Media Bus. [b7C+b7D]

[redacted] advised he obtained the following four registrations [redacted] [b7C+b7D]

491 DDD - Media Bus Incorporated, PO, Box 418, Lanesville, New York

[redacted]

62-0-13778

(1) - Albany
JSL/vep
(1)

NOV 11 1977
FBI - ALBANY

Buy U.S. Savings Bonds Regularly on the Payroll Savings Plan

The same FBI memo after lawyer Alan Sussman won an appeal to have the FBI declassify more information on Videofreex. *(Collection of the author)*

VIDEOFREEX

A flier announcing the services offered by Videofreex and Media Bus, showing how all roads lead to Maple Tree Farm. *(Media Bus archives)*

Maple Tree Farm REPORT

VIDEOFREEX   FALL 1971   ISSUE NO. 2

NEWS ON MEDIA BUS

MEDIA BUS, after a first summer on the road, contemplates what new autumnal craziness is in store for fall.

During the past 3 months the BUS has encountered many fine people, learned a lot, and made friends. Over 30 stops have been covered, including a womens lib conference, art shows, cable stations, a co-op, community action groups, a migrant workers camp, free school and a crafts fair. So, aside from the ever present equipment hassles, we feel pretty good about the ground we've covered, and encouraged from the positive responses received.

SOME RECENT STOPS:

★ **WOODSTOCK FREE SCHOOL**. We have been following the process by which a group (Earth Peoples Park) moves from the construction of domes and inflatables to house classrooms to the imrementation of these classrooms by children and teachers...

★ **AZTALON**...Ooo a pre-historic site in upstate New York showed video helping an anthropologist interpret and group artifacts found on the location...

★ **MADISON COUNTY CRAFTS FAIR**. At the 2 day fair, held in Oneida, over 100 early American crafts and 15,000 people were on hand.

Freex arrived with a full complement of hardware (Portapaks, monitors, 3600, etc.) which we set up on the porch of the museum. Over the course of 2 days we taped craftsmen making guns, candy, quilts, cabinets, wooden instruments, as well as the process by which raw wool becomes thread, how to throw a hatchet at your wife and miss, hoeing, corn grinding, and a fellow named **Cliff**, who built and knew everything about steam engines.

As the tapes were completed, we'd playback thru the two monitors to the craftsmen directly involved as well as to friends and passersby. In several instances we were able to tape the individual craftsmen, their reactions and raps, while watching tapes of themselves working. By doing this, not only did we document the crafts themselves, but we got a full swing (cause and effect) by taping the feedback of craftsmen reacting to themselves and their work, and commenting on both.

Our presence at the fair was one of a series of Media Bus stops that we have been making in association with **David Goff** and the **Madison County Historical Society** as a documentary process to preserve the older — and now rare — crafts on tape. (see workshops)

★ **EARLVILLE OPERA HOUSE**. In contrast to the heavy hardware stop of the **Crafts Fair**, we spent two days at Earlville, N.Y. without any equipment whatsoever. Earlville is a crossroads town with a population of around 1000. It has all the aspects of

"Smalltown, USA" and citizens take an active interest in preserving that aspect of its' identity to the extent of resisting financially tempting bids from industries wanting to move in.

Unlike most other smalltowns, however, Earlville boasts an opera house. Built in 1882, the opera house has seen minstral shows and movies but, ironically, no opera. Now, after 50 years of disuse, 6 artists and craftsmen (two of them local) gained title to the building as a non-profit organization called **FRIENDS OF THE OPERA HOUSE**. They are currently working to restore it and, when finished, the building will become a town center for events, workshops, and the housing of galleries and studios.

In order to raise funds for the costs involved in renovation, the Friends called up Media Bus in hopes that, as consultants, we would help them get publicity on cable stations thru video.

We began by suggesting they write up a package explaining the use of the Opera House and what the friends are doing. In addition we listed names of people and institutions where they could have access to video equipment, as well as calling our cable stations in the vicinity to find out the feasibility of viewing 1/2" tapes on the air. The Norwich cable station, "**VALLEY VIDEO**," was very receptive to programming as currently, they have no origination at all. We also suggested the N.Y.C. Educational Channel's program "**FREE TIME**" where 1/2" tapes have been used. We are remaining in close contact with the Friends of the Opera House to help insure them access to local media.

VIDEO WORKSHOPS

VIDEOFREEX IS GIVING A NUMBER OF WORKSHOPS THIS FALL EXPLORING THE USE AND RELATIONSHIP OF VIDEO TO DIFFERENT INTERESTS. WE ARE OPEN TO AND HOPEFUL OF EXPANDING THESE WORKSHOPS, PARTICULARLY ON A COMMUNITY LEVEL.
THEY INCLUDE:

1. **YALE UNIVERSITY POETRY SEMINAR**.

In association with poet Hugh Seidman, Freex is conducting a workshop on the use of visuals and voice-over. Although the presence of video in poetry is nothing new, we will be further exploring the relationship between the two and the development of a kind of spontaneity between language and sight. (November)

2. **GEORGE WASHINGTON HIGH SCHOOL ANNEX**.

In a program sponsor by Open Channel and participating with the teacher and a writer, we are attempting to introduce video into a ninth grade english class to see if between this collaboration kids will develop their own programming. Hopefully the results of their tapes will be aired on Manhattan Sterling Cable TV. (weekly for the semester)   (CONT. 2ND PAGE)

The second and last edition of the Maple Tree Farm Report. It was in demand, but most of us were too involved in making video to bother writing about it. *(Media Bus archives)*

8
Oh, Yoko!

A rtists, producers, documentarians, journalists, activists, actors, magicians, chefs, techies, bikers, politicians, hangers on, cops and spies, hundreds of them in the aggregate, visited Maple Tree Farm during the years we lived there. Some were colleagues like filmmaker DeeDee Halleck, who leavened her progressive politics with hearty laughter and who, like the Freex, had been seduced by video's promise as a democratic medium. Howie Gudstadt, Ben Levine and Ken Marsh from People's Video Theater rented a house in the next valley west one summer and beat Skip and me at badminton. Carol berated us macho country video boys for losing to the soft talking, ta'i chi-practicing city slickers. Cheatin' Lui showed up one day in his Cadillac, wearing a canary yellow suit. He hurried in, looked around like a cornered animal, smoked a few cigarettes to the half-way point and rushed off as abruptly as he arrived, saying he had to make the rounds of all the Catskills video groups and couldn't stay for dinner. It wasn't a far fetched excuse. At one time at least three other video groups, all smaller than ours, and numerous other individuals working in the medium, made their homes somewhere in these mountains. Or perhaps he shared Nam June's fear of the consequences of being mistaken for Japanese in rural America.

Bart's friend Ricky Jay sucked his thumb and held a blanket close to his cheek as he relaxed around the farm, but he transformed himself on camera into a poised, self confident performer, and remains to this day the world's most accomplished manipulator of playing cards, performing his magic off Broadway, on network television, in David Mamet movies, and once or twice on Lanesville TV.

When Bea Milwe walked into the kitchen, a diminutive, middle-aged, well-to-do matron from the Connecticut shore, I thought maybe she'd gotten lost on her way to visit relatives nearby. She opened her bag and pulled out tapes she'd made of a life in the People's Republic of China, on the street in Shanghai, a communal farm, an artist at home, the types of things mainstream media and video freaks had yet to discover.

A new group of Germans arrived, video people with their own portapaks. They looked at the grease-splattered sign on the wall above the stove listing tasks visitors could do to help out–suggestions other guests had long ignored–and went directly to the bathrooms and cleaned them. We had none of the tensions or misunderstandings with Charlie, Neidhardt, Marianna and Claus that we experienced with Valeska and her crew, a sure sign the video movement had taken hold in West Germany.

Members of the Ant Farm visited frequently after the conventions. They created some of the most striking art images of the era, including: *Cadillac Ranch*, in which one Cadillac from each of the ten years that car had tail fins was buried hood first in a row along the highway in Amarillo, Texas, with the rear half of the vehicles portruding from the ground as if the crests of successive waves of steel; *Media Burn*, which Skip helped document, in which a modified Cadillac goes crashing through a flaming wall of TV sets–aaaah, what a feeling of satisfaction; and *Eternal Frame,* a recreation in costume at Dealy Plaza, Dallas, of the sequence of events in the presidential limousine as Kennedy was assassinated, which passing tourists watched appreciatively as they snapped photos, assuming this was not guerrilla art but an officially sanctioned historical diorama. One Ant Farm couple mentioned when asked that, yes, they'd slept well at the farm, but they'd found the sheets in a guest room bed stuck together courtesy of some previous occupants. Arrgh! No one had bothered to change the linens. Recriminations flew as the horrified laughter died.

Lily Tomlin, an old friend of Nancy's, came for a working visit. She wanted to perfect a routine about a high school cheerleader. The best location Nancy could find for her was the plain brick wall at the Parish Hall in Tannersville. As we were setting up, with Tomlin already dressed in her cheerleader's short skirt and tight sweater, huge pom-poms in hand, the priest dropped by. Now what exactly is it you're doing? he asked. We tried to explain who Lily Tomlin was, how she was working on a bit for the NBC show *Saturday Night,* what kind of a show *Saturday Night* was, what we meant by a "bit." He looked more puzzled with each attempt and decided to stick around. The piece is a monolog in which Lily Tomlin becomes a sixteen-year-old talking about the things a teenage girl might be expected to, but with the twists that characterize Tomlin's extraordinary sense of humor. She uses a vernacular that sounds totally convincing, in part because it's laced with sacrilege: "Jeee-SUS," she says over and over to emphasize her points. "Oh m'god, oh m'god." The priest was not amused, but a higher power intervened before he could. We had an equipment failure and had to pack up for the day. He did not invite us back.

Each of these people and many, many more contributed to the vitality of life at Maple Tree Farm. Their tapes or their work in other media, or just their kibitzing around the kitchen table, affected what we produced, supplying new ideas and the type of independent, professional criticism our resident Lanesville TV audience couldn't give us. And on those nights when no more guests could squeeze around the table and the overflow had to lean against the cabinets or drag in a chair from the sun porch and look over someone else's shoulder, it seemed as if the video world had centered itself on the Farm.

The whole idea of video, though, was that it didn't and shouldn't have a single center. The evidence indicated the medium had quickly fostered a geographic dispersal of activity. The federal government of Canada had pumped large amounts of funding through National Film Board subsidies into video projects, from the cities by the U.S. border to the most remote Indian and Inuit villages. Cable companies were rapidly wiring cities and small communities around this country and Canada, and with cable came the possibility of access channels. People with portapaks in the Midwest, the South and West had independently begun to challenge what it meant to make TV. From Buffalo to Ithaca to Long Island, New York State Council on the Arts grants had spawned all sorts of video groups and boosted the careers of individual artists. But for all this activity, the true locus of the video movement remained in Manhattan. That's where the critical mass of artists and activists still chose to live, where the new people who wanted to make it knew they needed to come, where galleries would show the video work of the talented or well connected, and where even the relatively small number of cable viewers exceeded by factors of ten or more the largest audience most tapes were likely to find everywhere else. So I don't find it surprising that the person out of all our guests who exercised the strongest influence on my perceptions of the medium, and perhaps on the perceptions of other Freex, too, was a quintessential city person, who never felt very comfortable at the farm during her two brief stays with us.

∞∞∞∞∞∞

Shirley Clarke had learned through the alternate media grapevine that we had a video projector, and she wanted to see it. I didn't have the slightest idea who she was and sulked in the control room after having to interrupt an editing session at our Prince Street loft to hook up the old gray beast we used to project our regular Friday night shows. Allen Sholem got all worked up when he heard her name. "She's a big time director," he said. His tone sounded fawning, and he couldn't have chosen a worse way to introduce her in 1970. Big time to me meant she represented success in the way the straight media defined it, and I had become doctrinaire enough to hold her success against her. David and Davidson showed her the machine and spoke to her. I went back to work, having confirmed my suspicions she would treat us with condescension. Whenever I looked up through the control room window, I could see a small, bony woman–I couldn't tell exactly how old–gaunt and wan and wearing too much make-up, which only emphasized her pale, pocked-marked complexion. I found her at first glance homely, an impression enhanced by her black pants and jacket and the black, Greek sailor's cap, which looked as if it was all that held her scarecrow salt and pepper hair to her head. Under each arm she carried a miniature poodle, one black, the other white.

Shirley was big time, but not in the way I imagined it. Her first career was as a dancer, but in the 1950s she began to make films, some of them dance films, others far more experimental. She explored the possibilities of what for film was the portable, inexpensive medium of sixteen millimeter long before it became fashionable, winning some critical recognition along the way. In 1960 she changed gears, making a movie of a play called *The Connection*, by her own account the first independent film directed by a woman to have a theatrical release in this country. Her next film, *Robert Frost: A Lover's Quarrel With the World*, brought her the 1962 Oscar for the best documentary of the year. She followed that achievement by directing an adaptation of a novel by Warren Miller called *The Cool World*, a movie that true to form broke new ground for her and the industry. *The Cool World* told the story of a youth gang in Harlem using a documentary style and a cast of non-professional actors. It intentionally blurred the distinction between improvisation around staged incidents and what happened naturally in front of the lens, and it did this in a way that invited the camera into the action rather than leaving it always on the outside looking in. *The Cool World* received widespread acclaim, and its pioneering technique made it one of the most influential films of its time. Other directors followed her lead to such an extent that her work influenced an entire new generation of video and filmmakers, including Videofreex, though most of us didn't know it at the time. Shirley latched on to video not long after half-inch gear came on the market, embracing the medium in a way that sometimes made the Videofreex lifestyle seem dilettantish and disengaged by comparison.

Shirley had inherited family money and she and Max and Morris, her skittish poodles, had taken up residence in the penthouse apartment of the Chelsea Hotel on 23rd Street in Manhattan, a building famous as the home of many artists. I can't recall now how I first came to visit her there, but I spent many hours in her rooftop space, usually at Shirley's invitation, and not always for any clear reason I could discern, though I believe she always started out with some specific goal in mind. Once I arrived after the science fiction writer Arthur C. Clark, also a Chelsea resident, had dropped off his portable laser. Bart, Shirley and Nancy had invented a nasty little game with it, which I readily joined in. The beam from a flashlight or car headlight quickly broadens and loses intensity the farther from it you stand. But not a laser. It produces a focused beam. Pointed from the roof, Clark's crimson laser projected an intensely bright spot the size of a quarter on the sidewalk; and no one but the most observant person on the street would have guessed it came from the roof of the Chelsea. To people on the sidewalk, it looked as if the spot itself glowed magically as it moved along the pavement like a fluorescent cockroach. Bart said he, Nancy and Shirley had used it to introduce people to each other and confound a cop. Unlike today's pen-sized pocket lasers, this laser was about the size of a video camera and none of us had ever seen one before; we were mostly ignorant about its potential for damaging the eyes, and it would be years before we'd hear that laser dots were used as a sighting device for high-powered rifles. I have no idea what Arthur Clark was doing with his laser. Perhaps it just intrigued him. And by the time I got to the roof, Shirley had already become bored with it and the game. Her interest in technology did not extend to such long-distance, anonymously manipulative relationships. She preferred to use tools at close range to strip away the secrets of human interaction.

Shirley had little patience for the tools of the medium. She treated video equipment as decorative furniture. She adorned her monitors, cameras and VTRs with nail polish and glitter after painting most of them a matt black, including the pieces that came from the factory in black cases. Shirley was an artist, but she was not a painter, and every physical thing she did had a sloppy, hasty quality. Paint dripped onto screens and lenses, glitter floated into places it shouldn't have. She seemed not to notice. I suspect this amounted to a statement about the secondary role the equipment should play in any artistic undertaking. She didn't respect it, and quite frequently, it paid her back in kind.

Shortly after we moved to Lanesville, Shirley asked me to participate in one of her projects. She had a concept for a multi-camera, multi-monitor dance piece set on the roof of Westbeth in lower Manhattan, the new name of the old Bell Labs building at the corner of West and Bethune streets. After AT&T moved the labs to New Jersey, the building had been converted to apartments for artists. I knew it pretty well because I had once tried to hook into the internal cable TV system of the building so some video artists living there could cablecast tapes to their neighbors. That experiment hadn't worked, but it had whetted my appetite for unauthorized transmission of video to a receptive audience.

Shirley hadn't mentioned that she planned to shoot this project not just on the roof of Westbeth, but on the water tower on the roof. Before I could protest, she had me strapped into the camera brace and was instructing me how to climb into the scaffolding that supported the old wooden tank. She expected me to videotape a dancer, also in the scaffolding, while a crew on the roof fed me enough cable to allow me some freedom of movement–sort of like a deep sea diver in reverse. Meanwhile, she would make a separate tape of me taping the dancer.

I don't like heights, but it was too late. The crew had already attached the cables to me with gaffers tape. And then came the last, unannounced part of the shoot: the masks. Shirley instructed everyone on the set to wear Halloween masks which she produced from a large sack at her feet. The few people brave enough to ask why the masks were necessary received one of her trademark pained smiles. I refused my mask on safety grounds, although I might have been better off wearing one. It would have restricted my view of the street a dozen stories below.

As I began my ascent of the rusty scaffolding, I couldn't remember ever willingly having put myself in such danger. Shirley hadn't offered to pay me, but I hadn't asked because money wasn't the point. I never would have done a stunt like this for anyone but Shirley. On our next grant application, Media Bus could claim credit for serving the citizens of New York with this rooftop "workshop." I had become a publicly funded daredevil.

The dancer, a lithe Asian-American woman about my age, gave no indication this gig called for anything out of the ordinary. But what did she have to worry about? She had both hands free, while I was expected to use my hands to operate the camera lens. For what felt like hours I chased that dancer around the scaffolding. Below, Shirley occasionally barked directions, which became progressively more bizarre and hard to understand. When I finally made myself look down, I saw she was busy with the masked crew recreating a scene from Federico Fellini's movie *The Clowns*. The dancer and I were totally forgotten. I scrambled down from the tower and threw off the harness, finished with Shirley forever.

Shirley showed up a few weeks later for the wedding of Cy and Curtis, left her mark by shooting her unwelcome "comedy" and then retreated to the Chelsea, leaving behind some hard feelings. So it came as something of a surprise when she invited me and Carol to accompany her to John Lennon's birthday party. Shirley had known Yoko Ono for years and had received an invitation to the event. It combined a birthday party with the opening of Ono's show at the Everson Museum of Art in Syracuse. Shirley wanted us to come to the city and fly with her to Syracuse.

The thing that attracted Shirley to me was not my artistic vision. She wanted my technical know-how. I couldn't hold a candle to Chuck as far as repairing equipment or designing circuits, but I could fix rudimentary problems and I could explain what I was doing in a way that others could understand, even Shirley. I also knew how pieces of equipment got wired together so they'd work, and that, combined with my skills as a camera operator, was why Shirley wanted me along on the trip to the Everson. She said the invitation called it a costume party, and she planned to wire herself up with a portapak and camera that would transmit the image and sound from the camera to two round, glitter-encrusted TV sets operated by batteries and lugged by me and Carol.

We arrived at the Chelsea the night before, and I could see right away that Shirley's system, which I had suggested to her in an offhand way some time back, wouldn't work. I did what I could to patch it together and tried to give her the bad news as gently as possible. She refused to hear it. She had an abiding faith in the ability of her will to prevail over recalcitrant technology. I didn't argue.

A full-sized tour bus pulled up at the hotel the next morning. Neither Lennon nor Ono knew many people in Syracuse, so they rented a jet liner to fly in a few hundred friends and celebrities from New York and put them up overnight at a local hotel. Among the people who boarded the plane were Nam June, Abbie Hoffman and Jerry Rubin and Holly Woodlawn, a transvestite who worked with Andy Warhol, and who complained about his nails and lipstick during the whole trip.

The opening had all the trappings of a glamorous and exclusive affair, but where were the guests of honor? I admired Lennon and hoped we'd meet him, though I had nothing in particular to say. I felt extremely silly dressed as Shirley's electronic handmaiden, wired to her and to Carol, because the plan to broadcast from the portapak to the monitors worked only in theory. Fortunately, Abbie wasn't talking to me, so I didn't have to explain yet another broadcast failure. I looked around the room and realized no one else had come in costume.

Shirley wanted to see Lennon and Ono in the worst possible way. She became increasingly agitated as the evening wore on and the two did not appear. She was rude to some of the other guests, and because we were physically connected, we couldn't put any distance between us and her. Then people began to whisper that they had arrived and were showing a few select friends some of Ono's installations. Shirley sprang into action. Commanding Carol and me to follow her, as if we had much choice, she pushed past some security guards and headed out into the main hall of the museum. Across the room was an entourage. It was too dark to see any faces, and everyone in that group moved quickly and in the same direction. "Oh, Yoko!" called Shirley, waving her free hand in the air and tugging Carol and me behind. "Oh, YOOOOKOOOO!" She tried to make her voice sound musical but it came out strained and desperate. A door opened at the other side of the room, an exit, and the entourage ducked out. Shirley stopped and slumped. She seemed near tears. She'd done all of this for them and they hadn't seen it, hadn't even looked back, didn't care.

I looked around the museum at what I could see of Ono's pieces. I found them static and lifeless. Shirley, sitting on the cold floor, silent and forlorn, glowed in the light from the monitors we carried. We were a TV universe of our own, a complete network, a closed circuit that walked and talked and could grow and change as others joined us. The metaphor had escaped me in the hassle of patching together her harebrained system. She used video as an organic force in a way that intentionally scaled down the technology, diminishing its importance to the point you could approach it naturally, playfully. Creativity ruled the tools, not the other way around. No one had done this before, and now I understood why she clamored for the attention of people who had become icons: What interested them, by default, became the interests of a huge public that might benefit from a shake-up of popular perceptions of TV. Patrons like Lennon and Ono could have given her a bigger audience, assuming she could find a way to let them know what she was trying to do. Millions of people knew video only as television, and television only as its passive receptors. Shirley, more than all the video groups and all the studiously academic video artists, possessed a truly revolutionary vision for turning television on its head. Like Lennon and Ono, it seemed no one in the medium would stop long enough to notice what she'd accomplished.

∞∞∞∞∞∞

Shirley came back to the Farm one more time. She brought Max and Morris with her. They didn't get along with Mushroom or our cats. Maybe because of that, she decided to take up residence in the most remote room at the farm, at the very rear of the little house connected to the main structure by the second story bridge. She arrived during a prolonged period of cold drizzle. Nothing to do outside and no one had a major project under way in our studio. To keep herself amused, Shirley decided to decorate her room.

I think she might have had the fantasy this would become her home in the country, one dingy, unheated room with an especially low ceiling and peeling paint on the walls. She decided to dress it up as a French bordello and proceeded to buy crimson wallpaper with red velvet stripes. She painted the ceiling a deep red and hung the wallpaper with the help of Annie. But the job proved too much for her. She couldn't persuade anyone to join her fantasy. I can't imagine what sort of a scene she had in mind for that room; she was not a lascivious person and none of her tapes I ever saw ventured into nudity let alone sex. Whatever her concept, she broke down in tears, gave up just before it was done and rushed back to the city, leaving us with a rarely used room we called the Chambre d'Amour. It looked a lot like one of her decorated monitors, only in red not black. When people asked how the room got its name, we would roll our eyes and say it was too complicated to explain.

9
What the Clowns Are

Skip and I caught up with the circus in Fleischmanns, about fifteen miles west of Phoenicia. The bright August day had grown hotter all afternoon despite the breeze. Not the best shooting conditions. We arrived at the village recreation field to find the Royal Wild West Circus set up in the open, with bleachers facing one side of the two rings. Rigging for the high-wire act sagged overhead. Through earphones I could hear the rattle of a diesel generator parked off by the backstop. The field was deserted, so we picked out ringside seats with the sun at our backs and set up the camera.

A small man with sandy hair rolled a wooden stool for the elephant act to the edge of the ring. His short-sleeve coveralls, loose as they were, couldn't disguise the hump on his back. I asked who I should check with for permission to tape. He pointed to a trailer behind the stands. "Oswald Royal," he drawled. "He's the boss."

I knocked, and a grim faced man in his mid-fifties opened the trailer door. Royal–his non-circus surname was Schleentz–stood about five-foot-five, with thick shoulders and a square face. Even in the deep shadows of the doorway I could see his skin had tanned to leather. He wore a fancy white cowboy shirt, scalloped pocket flaps and embroidery on the breast, and his pants were tucked into the tops of tooled cowboy boots. He told me sure, tape all you want, just stay out of the way. No questions. He just shut the door.

When I was four, a big tent show came to the field across the street from my grandparents' house. The elephants tethered 30 feet from the side porch loomed bigger than a steam locomotive, loud and mysterious. One night they disappeared. Nothing left but trampled grass, flyaway paper and elephant dung, which Mr. Fowler, my grandparents' fussy neighbor, hurriedly carted off to his rose garden. A miniature of that circus had materialized here in Fleischmanns: one baby elephant, wire walkers head-high above the ground, ponies, not horses, prancing the ring; and all the performers would have fit in a clown car. Everyone did two, three or four acts.

Viewed close-up, the circus surrendered something of its illusion but none of its mystique. We had not gone to Fleischmanns to launch a major project, just a probe. So it didn't matter that our batteries ran low before the show ended or that intense sunlight overwhelmed the camera's imaging circuitry, rendering some sequences into silhouette. We got what we went for.

We sat next to the three-piece band, an organist, a drummer and a trumpet player, who doubled as ringmaster/intermission singer and a magician. Above the music I could hear adult carping. Better to see it at Madison Square Garden where it's never hot. Where's the tent, anyway? What's a circus without a tent? I couldn't connect with the complaints. I was hooked on the circus.

∞∞∞∞∞∞∞

A mixed crowd of local residents and summer people filled the lower rungs of the stands at the Royal Wild West performance in Fleischmanns. It surprised me to see so many Hassidim among them. I wasn't accustomed to Hassidim in this part of the Catskills. In our house-hunting phase we had considered a place in South Fallsburg in Sullivan County. It was too small, too close to the road, but David pressed us not to reject it so quickly. He liked that it was in the heart of the Jewish Catskills, the Borscht Belt. Close to the city, he said, and a nice place to be. I knew the area from my childhood summers in exclusionary communities in the nearby Town of Bethel, an Old Testament name adopted by the original Christian settlers. My grandparents referred quietly to "them," the strange men wearing long beards and dark suits even in summer, speaking their own language, heedless of the impression their brusque manners made. Not like Jews, like foreigners. "Now we know," my grandmother said, "why Hitler kicked them out of Germany." Such casual bigotry infused my childhood.

Fleischmanns, 40 miles north of the Borscht Belt across the highest of the Catskills, had been the mountain village of Griffin's Corners, a turnpike stop and a sleepy railway station with a gentile populace until 1883. That year Charles F. Fleischmann, a wealthy yeast and liquor merchant turned politician, decided to make the village his family's summer enclave. The Fleischmanns arrived just as the surrounding region, most notably hotels, succumbed to unconcealed anti-Semitism. In Fleischmanns, at least, Jews could find a welcome. The Fleischmanns were Jewish.

Sam had no use for Hassidim–"crazy religion guys," he called them. But he harbored no illusions about local anti-Semitism. He believed the German-American Bund, the Nazi front organization, used to meet at the farm, and I accepted what he said because I was prepared to believe it. The nearby Town of Shandaken has a large German-American population. I didn't know most of these families had arrived well before the Second World War, some before the end of the American Revolution, and displayed no sympathy for Nazis. We saw a crude election poster for a racist, far-right party on a telephone pole near Phoenicia during the '72 presidential election, and with our fixed beliefs about the poor, white, Christian population in and around Lanesville, we concluded this must be an inhospitable area for Jews.

From the mid-1820s, when the Catskill Mountain House was built a dozen miles north and east of Lanesville, and lasting through the First World War, the Catskills were the premier vacation destination in the whole United States. Industrial expansion in the nineteenth century created a new class of people who could afford to take a vacation, and brand new railroads brought them here. Some hotels and boarding houses advertised a "no-Hebrews" policy, but bigotry's grip was not universal. Jews and gentiles mixed freely at many establishments, while business boomed and the number of hotels and boarding houses in the region grew rapidly.

Author Stefan Kanfer in his book, *A Summer World*, describes how Boris Thomashevsky, the idol of the American Yiddish theater, built huge indoor and outdoor theaters on his estate in Hunter around the turn of the century, attracting an audience from a "substantial [local] Jewish population." Thomashevsky performed there, as did vaudeville acts from the city. He also showed movies, and while he may not have introduced the medium to the Catskills, he was probably the first to offer a local interactive performance the day he tossed rocks at his own image on the screen.

In Tannersville in 1917, the Ladies Waist and Dressmakers' Union, Local 25, of the International Ladies Garment Workers Union, announced the opening of a 500-acre union resort called Unity House. The union had grown in strength and resources following the horrific 1911 fire at the Triangle Waist Company in Manhattan. One hundred forty-six young women and girls, most of them Jewish, died in that fire, many after they jumped from the top three stories of the ten story building, where they were trapped by the flames behind bolted sweatshop doors.

Unity House established the first political resort in the Catskills, at least political in a way we would have felt comfortable with; yet we knew nothing of it. According to historian Alf Evers, the Communist Party gathered secretly in Woodstock in 1921, and we didn't know that, either. Nothing left when we arrived pointed to unconventional, let alone radical tradition other than the arts in Woodstock.

And yet we didn't know that Ralph Whitehead had chosen Woodstock for the site of his Byrdcliffe Art Colony precisely because the town didn't have any Jews. Woodstock had changed considerably by the time we came to the area, in ways that would have appalled that old English anti-Semite. The civil rights struggles of the 1960s had helped drive the most overt forms of religious hatred underground.

The 1890 census listed ten boarding houses in Lanesville, four up the mountain at Edgewood and a total of one hundred fifty-five in Hunter, Tannersville and Haines Falls combined. A decade or so after that count a Lanesville man named David Crosby built himself a three-story boarding house he called Echo Cottage. He nailed scalloped shakes around the third floor dormers and built a porch with milled woodwork, probably from the J.V. Neal & Sons chair factory up the road in Edgewood. The porch ran across the front and wrapped around one side, where it connected to an enclosed sun room. The front entrance was a double door, making the exterior a little grander than the narrow, dark hallway just inside. He built a small barn for the cows and a chicken coop in the side of the mountain. Livestock was necessary to feed the guests–gentiles only.

Progressive reformer Alice Hyman Rhine observed in 1887 that the boarding houses of the region were the worst offenders when it came to anti-Semitism, a phenomenon she attributed to intimate living quarters and to guests insecure about their own position in society. As the boarding house business declined after World War I, economic necessity overwhelmed old prejudices. The gentiles-only policy at Echo Cottage changed after David Crosby's death. Ruth Latz, who grew up and was married at Echo Cottage, said her mother relaxed the rules. German guests and "a lot of very, very nice Jewish clientele" stayed at the place in the summers of the 1930s and '40s. No one in Lanesville knows anything about the Bund anywhere in the valley, let alone at Echo Cottage, which Sam and Miriam Ginsberg renamed Maple Tree Farm.

∞∞∞∞∞∞

Sam and Miriam's decision to rent Maple Tree Farm to Media Bus left only one functioning boarding house in Lanesville in 1971. It belonged to a man known locally as "Rabbi Kelly" and his wife, Bertha. I first encountered them under less than ideal circumstances.

When we arrived at the Farm, every upstairs room had multiple beds, sometimes three to a room. We stuffed as many bedsteads and mattresses as we could in what was to become Shirley's Chambre d'Amour. They spilled over into the old shower stalls in the bridge, and there still wasn't enough room for everything we wanted to get rid of. So Sam arranged for us to store some of the bedding in the barn of Rabbi Kelly, who happened to be his brother-in-law. "Rabbi Kelly's" real name was Josef Keley, and he wasn't a rabbi; he had been a cantor. People in Lanesville called him rabbi because he always wore a yarmulke, a practice that struck them as exotic. They pronounced his name "Kelly" because it sounded familiar, and common usage had long since submerged any sense of irony at its Irish Catholic sound.

Joe Keley was born in Hungary and moved to Brooklyn before the Holocaust. He used to follow Bertha Ginsberg home each day to ask if he could read her newspaper when she finished it. He really did want the newspaper, but he was single and so was she, and with both of them lacking better prospects, they married. Joe and Bertha came to Lanesville before Sam, and bought a house even bigger than the Farm at the center of the hamlet. Joe had alerted Sam that Echo Cottage was for sale, but there was no love lost between the two men. "Oh, dis guy," said Sam more than once, angry at the mention of his brother-in-law's name. "Dis guy is de stingiest man in de woild."

Joe was smaller than Sam, just a little over five feet tall, slight and wiry in a way that made me wonder if he'd once been an athlete. He always wore a white shirt and dress pants. He shaved infrequently, leaving stubble but never a full beard. He used broad, violent, theatrical gestures, and when he chose to smile, which wasn't often, you could glimpse a devilish sense of humor. As far as I could tell, Joe Keley had never abandoned the mannerisms of the shtetl. He spoke in commands and demanding questions in a voice that sounded slightly strained and at a distance, as if it emanated from somewhere far down in his torso.

From the moment we arrived with our load of furniture he began hectoring Sam and demanding of us: "You know anybody who wants to buy this house? It's a great house and

I'm selling cheap because I'm telling you the tenants I got now they should be in prison for what they owe me in back rent and if you harm anything in that barn I'm going to make you pay double, you hear me, double!" Sam cursed him in Yiddish and that precipitated a shouting match between the two men that lasted the entire time it took us to lug the furniture into the barn.

Joe and Bertha's boarding house qualified as seedy even by Lanesville standards. It needed paint, evergreens draped over the roof, and an abandoned junker or two usually graced the driveway. A few of his old clientele, orthodox Jews from Brooklyn, still came up for vacations. They provided the only outward signs of Jewish life in Lanesville as they settled in on the front lawn, sharing the patchy grass uneasily with the hippies and musicians and other Keley tenants harder to classify.

Joe would rent a room to anybody, even people with no obvious means of support. He treated all his tenants with equal disdain and relied on his abrasive personality to coerce rent collections. The only African-American I ever knew who lived in Lanesville, a flamboyant leathersmith named Jonesy, rented a room at Keley's. He complained bitterly of Joe's penurious behavior, but never of racism.

∞∞∞∞∞∞

The Royal Wild West Circus spent only one day in Fleischmanns. It packed up a few hours after we left and moved to Cairo–the natives call it KAY-roh–about 50 miles to the northeast in Greene County. I had already decided I had to follow the show, and I was happy to be finished shooting in Fleischmanns. The previous summer David was heading back to Lanesville from a mid-state workshop when he came to a state trooper roadblock near the village. He'd been blowing a joint and was too stoned to make sure the car was clean. They nailed him for something, a roach in the ashtray, ashes on his clothes. We found him a young Woodstock lawyer named Robbie Ricken, a former assistant district attorney in the Bronx. Robbie hunted up the town justice, Solly Darling, and found him at his laundromat.

Darling evinced little tolerance for hippie dopesters at a time when Rockefeller's new, draconian drug laws had just taken effect. He controlled the process that could have started David along a route that led to years in state prison. Robbie assessed the situation and announced a series of motions he planned to file on behalf of his client, informing the judge along the way that David was a gainfully employed artist who resided in another county and had no plans ever to set foot again in the judge's jurisdiction.

Robbie challenged the legality of the search in a way that promised a protracted trial costing far more than the normal expense involved with making an example of long-haired malefactors. Darling sized up Robbie, weighed the possibilities and rendered his judgment: a fine to be paid on the spot and a promise the defendant would not bring his despicable self to town again. David, a bit shaken, was a free man.

We greeted the outcome with relief rather than jubilation. Nobody wanted him to go to jail, but we suspected David brought the bust on himself by acting with characteristic obliviousness to his surroundings. Some of us felt he'd put the whole farm in jeopardy.

In the back of everyone's mind was the fear that the cops would seize any excuse to raid us. I was coming down the stairs one afternoon when a Freex stampede heading the opposite direction nearly knocked me over. "There's a cop coming," whispered Chuck on his way by. His voice was tight with fear. Annie clunked up the stairs behind him. Doors along the hallway slammed shut. I walked into the kitchen, where one of our guests stood sheepishly with a local constable. He had forgotten his car registration and the cop, courteously, followed him back to the farm so he could prove he had it.

Another time, a constable wanting favorable coverage from Lanesville TV pulled me aside to say he'd looked the other way as I removed Nancy's marijuana plant from the bathroom window during another unannounced visit. False alarms, sure. But none of us trusted anyone with a badge to protect the rights of a longhair.

∞∞∞∞∞∞∞

Cairo was hotter than Fleischmanns. The humidity had shot up. Dark clouds flattened the light, making for better images in the camera. I hung out with the performers behind the bleachers. No one mentioned my long hair. For the first time in years, I lost my hippie identity completely among people who, except for their costumes, looked very straight.

Eddie Henderson, a thin man with wavy hair, was a wire walker who had fallen too many times. Now he had a dog act. Like almost everyone else in the circus, he naturally played to the camera, friendly and open. He takes a tug at his bourbon and Coke in a paper cup before entering the ring. He sipped from that cup all day long and must have been looped, but he doesn't miss a trick. His family consisted of his dogs, almost a dozen miniature poodles. He cooed over them constantly and carried on long conversations with his favorites.

Ossie worked the elephant. He stood a head taller than the animal and prodded it gently with a pike as it performed. In Cairo the elephant balks, frightened perhaps of the lightning now visible in the distance. He can't get her to budge.

"See what the clowns are? They are the glue the circus is made of," intones Joe. Leaning his red and white face toward the camera, he sighs, "The act isn't ready so the clowns fill in."

Joe turned seventy that summer. He was a cop in a small Pennsylvania town who ran away to join the circus. He wouldn't say if the story had more to it than that.

No one asked me a personal question. Circus people don't pry into your past. Breech their decorum and you'll find yourself treated like an outsider, a rube. So I restricted myself to observing not questioning, which gave me the freedom to peer pretty much wherever I chose.

The rain held off until after the first show, when a thunderstorm rolled over the edge of the eastern escarpment of the Catskills and whacked Cairo for what seemed like an hour or more. Ossie strode into the American Legion Hall wearing a yellow slicker. In the field outside hail pounded the trucks and trailers of the circus. A few locals were drinking at the bar, and he checked to see none of his people were among them. Canceling the second show meant losing money, but it was too late to raise canvas in time for the evening show. "Looks like we might see a ... ." He swirled his forefinger in a funnel. Jimmie, his crew chief, the man with the hump, nodded. A tornado can collapse a tent in seconds. It's every mud show's nightmare.

Ossie decided to keep the circus in Cairo that night and drive to Phoenicia early the next morning. Each day they set up, did a show or two, broke down and packed, and drove fifty to a hundred miles to do it all over again, repeating the cycle for weeks on end without a break. Performances started in early spring in Georgia, where the Schleentz family lived. They played as far north as Maine, and worked their way back to Georgia by late fall.

The storm rumbled off at dusk. Under floodlights, the crew and performers went about the business of stowing all the gear on the trucks. The camera captures an occasional ghost-like image, the glint of a face. A group of performers runs toward the marquee. I hear shouting and follow, jogging as fast as I dare with the portapak on my shoulder. You could never tape and run at the same time with a portapak. Changes in momentum played havoc with tape speed and contact with the heads. Even when shooting while walking, you had to move smoothly, like a dancer.

I ease my way to the front of the knot of circus people. In the headlights of the forklift, Ossie's son, Don, faces off in ankle-deep mud with one of the Flying Rosarios, a family of aerialists from Columbia. The aerialist, Francisco, has a knife. Don holds a tent peg. Don has the build of a weight lifter. He's six feet tall, with the all-American beefcake good looks of a '50s cowboy star. He does a trick shooting and riding act with his wife and little daughter, Princess Sharon. Rosario has a wedge for a body, shapely legs and huge shoulders.

The two men sway back and forth in a tense standoff. I fool with the camera, trying to find some way to pick up the scene. In the viewfinder all I have is the faint outline of Don's t-shirt. Ossie walks between the two men, puts his hands on his hips and tosses his head toward Rossario. "Go to my trailer," he barks. "You're done. We'll settle up." My batteries are dead. The sound is drowned out by the generator. The drama is unrecoverable from grainy streaks, mumbles above the din, tape dropout. Everybody else has gone back to work.

I spent the night in the trailer of Bob, the magician/musician. Bald and serious, he looked like a high school science teacher. Bob's parents lived in Pine Hill, a dozen miles west of Phoenicia. In the ring, he made Don's pretty wife disappear. He said I should refer to him as an illusionist, not a magician. A magician does parlor tricks; an illusionist makes crowds believe they've seen what they haven't. I tried hard to grasp this distinction as I fell asleep.

Before dawn the next morning, the show rolled south and west to Phoenicia, retracing much of the route it had taken from Fleischmanns to Cairo. Carol and Skip had brought me fresh batteries. Skip would tape the first trucks as they arrived in Phoenicia. I rode with Jimmie. The route was marked by small paper signs at key points. Each sign bore a red stenciled tepee, Royal Wild West's signature. The advance man had taped or stapled them at every turn in the route, with the apex of the teepee pointing the way to go. All road shows have their own distinctive route signs, and I used to watch for them each year to see who was passing through the area. Throughout the 1970s, all sorts of traveling shows, carnivals and circuses criss-crossed the Catskills, leaving their telltale markers behind. Since then, the number of shows has diminished judging from the absence of route signs.

The storm had brought with it slightly cooler temperatures, so Ossie called for the tent in Phoenicia. Raising the canvas required everyone in the circus, roustabouts and performers, and the bigtop filled much of the parish field. The Flying Rossarios had left, but relatives of theirs who had remained with the show, the Fernandez Family, direct from the great stages of Europe and Latin America, filled in with an extra balancing act. Neither Ossie nor Don said a word about the incident in public. Jimmie said it had to do with money.

I thought of myself as physically fit, but a couple of days with the circus exhausted me. The Royal Wild West show was ready to move on to the next lot, performing with the same showmanship as anywhere else so the sponsors would invite them back next year. Shorthanded, the circus forced men to double up-to drive a truck to the next town and be driven back so they could drive a second truck to the new lot. I offered my services. Maybe I shouldn't think about taping at all, just observe and tape later. But we had Sarah now, a little more than a year old, and Carol showed no interest in joining the circus. Jimmie laughed when I complained how sleepy I was. Ossie didn't respond to my offer.

The circus headed far upstate after Phoenicia. I didn't tag along. I wanted to do something with the tapes other than play them for Lanesville TV and then stick them on the shelf, forgetting about them or consigning them to be erased and recycled. I drove back to Maple Tree Farm from the last show in Phoenicia thinking about a circus series but was distracted by police cars and a small crowd at the general store. When I got to the farm, Nancy said she'd heard someone had been shot.

10
We Don't Mean You!

The Saturday of Labor Day weekend, 1973, at around 7:30 in the evening, a 23-year-old tenant of Joe Keley's named Thomas Musso pointed his .22 caliber rifle across Keley's lawn toward the General Store. Charlie Benjamin, Jr. was playing on the pavement in front of the gas pumps. He was 12 at the time. Musso later claimed he didn't see the boy when he squeezed the trigger and shot Charlie, Jr. in the stomach.

I must have counted half a dozen police cars–state, local and county sheriff's–at the store as I drove by on my way home from the circus. We seldom saw any cops in the valley. We liked it that way and it seemed everyone else did, too. There wasn't any violent crime in Lanesville that we knew of, and the infrequent appearance of police cars in the neighborhood became a matter for conversation at dinner. The people who smoked grass always got a little paranoid at the sight of cops, especially after David's bust, but by our third summer in Lanesville, all of us had relaxed about the police on the tested assumption that they had little interest in patrolling our side of the notch. What would they achieve if they did? They could always bust speeding skiers closer to Hunter. Maybe they'd pick up a few drunks or a deer jacker over here, but these cases weren't worth showing up for in court, especially if a distant relative or an old acquaintance might be the defendant. If they were coming after us, they'd probably have come before this. So a large contingent of police indicated something out of the ordinary had happened.

In addition to the cops, private cars, pick-ups and vans were parked along the road near the General Store that evening. It amounted to the largest crowd I'd ever seen at the store. A party out of hand or something like that, I figured.

Charlie, Sr. had purchased the store from Denny not long before, and we were still trying to get used to his style and that of his wife, Shirley, who sat behind the counter and did business with us, if you could call it that. Her manner bordered on comatose. "D'you have any peanut butter?" I'd ask. She meet that with silence for a few moments as if she wished to contemplate an answer. Then, "No-puh," a little, two syllable negative ending with a plosive that communicated not only the information requested, but the clear message that she would volunteer nothing more, particularly not casual conversation. I never took offense at her lack of courtesy. Her reticence suited me better than the hostile challenges from Denny, and I assumed she treated everyone else much the same. The meager stocks on the shelves served better than rumor to confirm the new owners intended to sell off what they could of the inventory and not replace it. When that was gone, we'd have no general store, nor a store of any kind in Lanesville. Too bad, we said. An unhelpful store was better than no store at all.

The distant nature of our relationship with Charlie's family meant none of us at the Farm felt comfortable inquiring about the reason for the police at the store. Soon enough, though, some of the neighborhood kids showed up with the story. John Benjamin told us his cousin had been shot by a hippie, with the stress on the hippie part. John didn't have anything against hippies, and if he considered us hippies instead of a separate class of "Videos," it certainly didn't involve disdain, judging from his readiness to continue in his role as Sheriff John in *The Buckaroo Bart Show* on Lanesville TV. The shooting of Charlie, Jr., however, constituted an assault on the family, and what he said reflected the topic under discussion by the adults around him.

We also learned that the police had arrested Keley. The kids said the cops took him because the hippie lived there.

We stayed away from the store the next day, Sunday. If what John and his brother Scotty said was true, we didn't want to inflame the situation. The kids returned later with the news that if young Charlie died it would mean big trouble, with all the hippies run out of town. They didn't say whether that included us, and we didn't put the question to them. Nor did they identify who had made the threat or what plans were afoot to carry it out. It didn't matter. We could sense the tension in the hamlet even from our vantage point a quarter mile up the road.

The shooting was big news by Lanesville standards, perhaps the biggest ever, and an opportunity for us to show that we could serve the community as a news organization, on the scene with a portapak and on the air with a special broadcast about the incident. No one else, not the newspapers or the radio stations, would give this crime much notice unless the boy died, and by then the grapevine would have circulated more news than any reporter could possibly cram into a story. We had an exclusive, something only we could tell the community. We could have gotten out the news that day. But we didn't.

I could attribute our inaction to fear alone, concern that the Benjamins–hell, the cops, too–would lump us in with whomever had done the shooting, seeing longhairs as longhairs, whether or not we carried cameras. More than fear, though, the reasons for our failure to react over the air lay with our ambivalence about the role of Lanesville TV.

∞∞∞∞∞∞

In 1973, television journalism had not yet discarded all pretense of civility, and some subjects still seemed inappropriate to broadcast, even by us. We didn't sit around the kitchen table wringing our hands in an attempt to rationalize our reluctance to intrude on the scene at the General Store. A socially conditioned response came into play: Be sensitive; respect the privacy of your neighbors. The Benjamins were in pain, so what right did we have to barge in on them for the sake of something called the news? We had peeked behind the scenes of TV news and knew how it flirted with entertainment: news as a product manufactured by the big money media. All of us Videofreex, regardless of how our views on the uses of video diverged, shared a skepticism of the mass media's methods of packaging information. On that weekend in Lanesville, we couldn't answer the question of how we could treat the shooting of Charlie Jr. without trivializing it.

None of us had committed to producing anything like a regular newscast on Channel 3. We presented Lanesville TV as our experiment. The community could accept what we offered or turn us off. No advertiser could sway us because we had none. No officials could regulate our activities because we operated without the knowledge of the FCC. We had redefined what it meant to broadcast to the community, or thought we had. If someone had called to ask us to cover the shooting, a couple of Videofreex would have grabbed a portapak and hightailed it down the road. No such luck. We failed to act because nothing compelled us to. We made up the rules as we went along.

It's a quaint and discarded notion now, but up until the 1980s, the federal government imposed a public service requirement on all broadcasters. Before the Reagan years, government policy acknowledged that the scarcity of broadcast channels meant anyone who had a TV station license possessed a virtual monopoly sanctioned by Washington. The cliche that a commercial TV broadcast license was tantamount to a license to print money held true for the majority of stations. So in exchange for these monopolies, Congress decided that those who received them ought to contribute to the public good with a few minutes of programming that responded to the needs of the community a station served. The regulations went so far as to state that a station could forfeit its license if it failed to provide adequate public affairs programming. Broadcasters responded by creating the "Sunday morning ghetto," talk shows with safe community leaders droning on about not-too-controversial topics early on a day when few people watched TV, anyway, and thus, when the broadcasters would lose little revenue by putting them on. The FCC, which could arrest J.P. and Allen for their violation of the rules, simply winked at the broadcasters' cynicism.

We had placed ourselves and our station outside this corrupt monopoly system by going on the air in defiance of the rules. So we didn't give the public service requirement a second thought. If we had, we would have said we didn't produce anything but public service programming.

The shooting confronted us with an event the public–our public–had a vital interest in knowing about, and what did we decide to do about it? We decided to censor ourselves. The fact that we ran a pirate television station did not lend us enough sense of purpose to overcome our fears. We felt suddenly exposed, uncertain what to expect next, no different, probably, from most of the people in Lanesville that weekend.

∞∞∞∞∞∞

None of us knew Musso, the guy who shot Charlie, Jr., but we'd met other tenants of Keley's, and knew a drug scene of some sort went on down there. That made it easier for us to make up our own explanations for the shooting, imaging Musso as a speed freak just off a long run of methamphetamines, wired and crazy enough to shoot at anything. In the absence of real evidence, it sounded plausible. Or maybe he'd been a set-up, the kid shot by someone else and a longhair blamed. Well, druggie or not, Musso had got himself in deep trouble, and when somebody from Keley's called, panicked, pleading for help, we offered Robbie Ricken's name as a lawyer adept at helping hippies in trouble. We couldn't do any more for Musso. But Joe Keley—why had police arrested him?

Nancy was having dinner one evening with John's parents, Harriet and Bobby Benjamin, at their trailer. John's grandfather Willie, a widower, who often crossed the street to have Harriet cook for him, was there, too, and he began to lace the conversation with anti-Semitic remarks, going on about the Jews this and the Jews that. No longer able to contain herself, Nancy stood up and told them, "I'm Jewish, you know."

"Aw, sit down, Nancy," they said. "We don't mean you!"

And Willie called during a show once to comment over the air on some aspect of Lanesville culture, telling Skip and the broadcast audience, "Those Jews, they going to stick together?"

Skip blew off the question with a non-confrontational, "Uh oh, Willie. I don't know what you're talking about," which was a lie. Skip must have felt too laid back to challenge the statement. Anyway, Willie was a regular viewer, and we welcomed calls from anyone no matter what the caller said.

Despite Willie's opinions on the subject of Jews, when the cops hauled Joe Keley off to jail on a trumped up charge of violating the health code by having too many tenants, but really because Musso, the guy who shot Charlie, Jr., lived with other longhair, dope-smoking young people without fear that Keley would kick them out for their lifestyle or appearance, and neither Keley's wife, Bertha, nor his in-laws, Sam and Miriam, knew how to drive, Willie hobbled out to his Cadillac on his two crutches and drove down to Catskill on Sunday to bail Joe out. Joe was a neighbor in trouble and needed help. Willie gave it and never made a point of what he'd done. He probably didn't mean Joe Keley, either, when it came to "those Jews."

Maybe the police arrested Keley to protect him from the wrath of his neighbors. His non-judgmental rental policies coupled with inexpensive rooms attracted unconventional people. Sometimes Joe would call the station during the show and holler about some deadbeat, probably because he figured the deadbeat had tuned in, too. Performances like that enhanced the local perception that the people who lived at Keley's constituted the dregs of humanity instead of people who couldn't afford the exorbitant rents charged by other landlords. So as people in Lanesville talked about the shooting, they began to view it as the inevitable outcome of allowing that kind of person to live in the neighborhood. Musso had surrendered to the police shortly after the shooting, and there was little likelihood he'd get out of jail quickly or that he'd return to his room at Keley's. But that left unresolved in the minds of many people in Lanesville the issue of what to do about the type of person he represented. They felt the incident demanded they do something.

On Sunday afternoon, we learned a community meeting had been called for that evening at the Methodist Church Hall. By that time, we'd heard plenty of rumblings about vigilante action and the anger directed at Keley. Even the darkest parts of what we heard sounded plausible. We'd been in Lanesville a couple of years and the station had been on the air nearly a year and a half, but that hadn't changed our status as outsiders who had little regular contact with most of our neighbors. Lanesville lay too close to the orbit of New York City to cast itself as a true backwoods settlement. Still, images from the movie *Easy Rider* of the biker heroes blown away by a redneck with a shotgun set paranoia to work. Would our neighbors take the law into their own hands? If they did, what could we do about it? We had no defense but to flee or to record them carrying out their vengeful plans.

Around the kitchen table we debated whether to attend the meeting. Some Freex said they didn't want to go in case the meeting turned ugly and the wrath was directed at us. I argued that we lived in Lanesville and had every bit as much right to be here as anyone else. Sarah was just a little over a year old, and I didn't want to think she was growing up in a place where kids got shot. No one should have the power to exclude us from the meeting or to frighten us into thinking we didn't belong there. And, as a practical matter, by showing up, I figured we'd make it harder for folks to lump us in with the people who stayed at Keley's. Any blanket condemnation would have to take us into consideration, and any plans for violence risked finding their way onto Lanesville TV, whenever we got around to reporting this story. So much for any lingering feelings of solidarity with the longhair hoi polloi–this was the first stirring of the burgher in me, the pirate broadcaster/bourgeois.

Television broadcasting can do this to a person, for minuscule though it was, our target audience amounted to the masses of Lanesville, and the longhairs were definitely not the masses. When it suited our needs, as it did now, we were part of the Lanesville proletariat, whether they had any intention of accepting us or not. My arguments prevailed in the sense that while we didn't go en masse, several of us, including Carol and Sarah in her stroller, showed up at the Methodist Church that evening at seven.

An angry buzz filled the room as people rehashed what they knew about the shooting. I half-expected the room to go silent when we walked in. My palms sweated and I could feel my heart beating. But nothing much happened, maybe an icy stare or two along with as many hellos while we took our seats and waited for the Benjamins to call the meeting to order.

The temperature of the crowd dropped noticeably with the news that doctors expected young Charlie to make a full recovery. But if the tension left the room, the sense of bitterness did not. Somebody called for the cops to close down Keley's house and kick out all the hippies. But somebody else, it might have been Jerry Doyle, jumped up to point out that if the cops found reasons to close Keley's place, they could just as easily apply the same standards to all the other rental units in Lanesville, and that would screw up the biggest industry in town. The wisdom of this analysis was not lost on the meeting, and the idea of inviting scrutiny of Lanesville landlords quickly lost steam.

The Benjamin elders, with the exception of Charlie, Sr., who was at the hospital, presided at the meeting, as taciturn as ever. That provided an exceptional opportunity for the other fifty or so people in the room to vent their feelings. And as there was nothing much more to say about Musso and Keley, the speakers soon turned to expressing frustrations about their second-class treatment by the town government run by men who lived and worked over the notch at the Hunter Town Hall in Tannersville. Issue after issue came to the surface: the need for stop signs at intersections with Route 214, a streetlight near the center of the hamlet and a speed limit, for a local phone number rather than one in another area code to call for emergency services, and for a town cop stationed in Lanesville–all of which comprised the normal types of requests communities make of their political leaders.

The prevailing view of the citizens of Lanesville held that it never did any good to ask a politician in Tannersville for something because Lanesville amounted to nothing more than a bunch of "woodchucks," and politicians didn't pay attention to woodchucks. But here, suddenly, was a significant segment of the population of the hamlet, whose members, perhaps for the first time, realized they had common interests, that they could look at themselves as a community. More important, they might just have the power to get what they wanted.

Beecher Smith, the owner of the sawmill, suggested a committee that would come up with a list of Lanesville grievances. That led to a discussion of who should be on the committee. At first, none of the Benjamins volunteered, but Scotty Stuart did, as well as several other people, including boarding house owner Scotty Sickler and his son, Tommy, a Shandaken constable. Willie said he might sit in on the sessions–nothing much went on in Lanesville that Willie didn't get involved in. Babe Santora, owner of Babe's Edgewood Bar, was appointed because she had spoken up at the meeting; and Beecher was chosen because of his stature as a wealthy man and an employer.

It sounded to me as if they had overlooked one important constituency. I couldn't catch a deep enough breath, but because I had already opened my mouth about having a local phone line, which everyone thought was a good idea, I spoke up, anyway. "What about the renters?" I gasped. "They're not represented."

A chorus of voices answered at once, "Oh yes there are! Jerry Doyle rents out apartments and Beecher, too."

I'd meant "tenants" not "renters." I just couldn't think of the right word. My tongue wouldn't work, either. I could see this committee coming up with a proposal for identification cards issued to all the longhairs in town when Jerry Doyle broke in.

"Naw. He means havin' somebody like him on d' committee. I think that's a good idea. So whyn't you be on it."

In the spirit of discovery that had gripped the room, no one rose to contest Doyle's authority to invite me to join the group. This new sense of unity rested on so fragile a base that any expression of dissent would have destroyed the accomplishments of the evening. The assembled citizens spoke with their silence. They had approved me as the final member of the Lanesville Citizens Committee, and with that we all quietly left the meeting hall. The gathering had lasted a little over two hours, but in that time, the tension level in Lanesville had dropped perceptibly.

∞∞∞∞∞∞∞

The committee met a few days later back at the church hall. Beecher sat at one end of a long table grinning. He was a tall, thin man with a large head, a prominent Adam's apple and sharp features. He spoke infrequently in a reedy voice. He owned the last sawmill in the valley and several homes, some of which he rented to his workers, others to anyone who could make the security deposit. "I had some of them hippies living in one of my houses tell me there was vigilantes in Lanesville looking for all the hippies," he said. "He was worried about it, and I didn't say anything to change his mind." He chuckled as did everyone else around the table but me.

After living with Videofreex for four years I had learned plenty about functioning in a group. I listened quietly at first, but not meekly, offering occasional suggestions while taking care to say how each of my thoughts built on ideas introduced by someone else. It took a while, but slowly, the members of the group relaxed, reasonably certain I wouldn't flip out in a drug crazed tirade nor would I try to make points because I had more education than the rest of them. They decided I should take notes of the proceedings and rephrase and edit what would become the a list of demands by the citizens of Lanesville.

The grudging acceptance I won as a member of the committee did not make me popular at Maple Tree Farm. The other Videofreex believed I'd sold out because one of the demands was for a town cop stationed in Lanesville. Why voluntarily end the isolation from police that the mountains provided? I countered that our neighbors wanted a cop here and they were going to ask for one whether or not we supported the request. It seemed like part of the democratic process to me. In order to get some things we all agreed were needed, like a local phone number for the town hall and ambulance, we might have to give up our fantasy of living in a world without police. But I didn't smoke grass, and though I still felt uneasy when the police showed up, I didn't see a local constable as an immediate threat. I allayed some misgivings by pointing out how unlikely it was that the town would ever agree to stationing a cop in Lanesville, no matter what the people called for.

Within a week the list had been drawn up, broadcast to the community over Lanesville TV and about sixty people jammed the town hall to demand action. The Town Board members reacted at first in stunned silence. They had never heard from Lanesville, en masse before. Now three score angry voters, organized and prepared with a written list of reasonable demands sat glowering at them. Once they regained their composure, the board members agreed to most of the demands immediately. They wanted to investigate the costs of the phone line, and they said they needed to think about a Lanesville constable, but they promised an answer soon.

A number of people from Lanesville shook their heads in disbelief. They had never before exercised raw political power. Woodchucks or not, they had raised their voices in unison and someone who mattered had paid attention. The stop sign went up the next week, and the street light not too long thereafter. Then the phone line went in so that anyone in Lanesville could summon emergency services with only a four-digit call–at that time you could ignore the three-digit prefix on local calls. And then, something that surprised me, though I should have seen it coming, Tommy Sickler got himself appointed Lanesville's cop and was allowed to bring his patrol car home with him. I had to accept the reprobation of the Videofreex for that one. Many of the group held me personally responsible for the arrival of a police presence in Lanesville. We never saw Tommy at Maple Tree Farm, and if he ever mounted a serious investigation of us, it never came to anything. There was really no reason for him to have hassled us. We were generally law-abiding citizens, and I'd helped get him his job.

We dutifully reported the progress of the political appeasement of the voters of Lanesville on Lanesville TV. We could have stretched that commitment to covering Town Board meetings with our cameras, but none of us had the time to shoot the tape and edit it on a regular basis. It seemed like drudgery and a bad use of our limited resources. A group of men sitting at a table talking about town business for hours on end each month just wouldn't make engaging video. And if we started to fool with the images electronically, we'd invite trouble we didn't need. Groups all over the country had begun to take advantage of the FCC's 1972 ruling on cable TV access, using local channels on newly strung cable TV systems to relay the unedited processes of government to any viewers willing to sit through them. But we didn't cast ourselves in that role. We had only a limited amount of time on the air, and we wanted to use it as the whim of the week struck us. From the start, Lanesville TV maintained a sense of unpredictability and fun, but it never evolved into a particularly utilitarian service.

One reason the politicians in Tannersville responded so quickly to the demands from Lanesville after the shooting of Charlie, Jr. was because several of them had to run for election that fall. With Lanesville suddenly on the political map, a lot of the candidates made a point of showing up at Maple Tree Farm. Each member of Videofreex had registered to vote, which meant that the ten of us comprised about one percent of the registered voters in the town, more than enough to decide a close election. We never gave any indication that we voted as a bloc, though certainly our political sentiments lay well to the left of center and we didn't care who knew it. Candidates had to ponder whether we represented a unified force and scratch their heads over what influence, if any, we had on the other voters of Lanesville.

Jerry Doyle trooped by with one of those candidates, a guy named John Glennon, a Democrat running for town supervisor. Jerry had toughed it out in the political wilderness for a long time, almost singlehandedly representing the loyal opposition in Hunter, a Republican stronghold like the rest of Greene County. But he sensed he had a Democratic winner in Glennon. "You're gonna like him, you're gonna like this guy. He's gonna make a great supervisor," said Doyle, rubbing his hands. Glennon stood right behind Jerry while he gushed, grinning and shaking his head. "He's a former FBI agent. You're gonna like him ..." Glennon reached around behind Doyle to shake my hand.

"A what?" Too late. I shook Glennon's hand. He towered over Doyle, who stood about five-foot-six and looked shorter because of his protruding belly. Glennon had to be six-four, trim but a little gawky, with a smooth, round face. He connected with the ease of a natural politician, a direct look and no fear of small talk. The leaves of the maples lining the driveway had just turned. The sun seemed to heat the colors. Doyle rattled on about how Glennon had just left the Bureau and moved back here to his old home town and how, oh boy, now the Democrats were going to take over town government. I tried to follow him while glancing at Glennon. He looked about 35. Why would he quit the FBI to live in Tannersville?

I told him a little about what we did—the tourist version, cleansed of politics and any references to unlicensed broadcasting, just our little "experimental TV station," which held enough truth to make it plausible. I didn't find him threatening, but I didn't trust him completely, either. We had nothing on the first floor that required hiding. I decided to invite him in to see the studio. He acted as if he'd rather get on his way. He followed politely, deposited a few of his "John Glennon for Supervisor" ballpoint pens and left.

Glennon won the election that November, and took the oath of office the following January. Some of the Lanesville requests on the list remained unsettled, and one afternoon in the winter of 1974 Skip, Carol and I drove to Town Hall to tape a discussion with Glennon and several members of the Lanesville Committee. The talk plodded along. By now, the anger in Lanesville had dissipated and the momentum for change had ebbed. We had decided to do this tape because we had nothing better in mind for that week's show.

This was Carol's tape and she acted as moderator of the discussion. While Skip and I set up the camera, lights and mic, I noticed Carol talking quietly at one side of the meeting room with Glennon. I couldn't hear what they said and I had work of my own to do. I forgot about it until, during a break for a tape change, she came over to me and whispered, "You know what John Glennon just told me? He said when he was in the FBI he investigated our house."

She said he didn't want to talk about it on camera. She'd tell me more later. And with that, we taped the second half of the discussion without any reference to his revelation.

The other Lanesville people had hung around after the taping, giving us no opportunity to question Glennon in private; and this was not something we wanted to talk to him about in front of our neighbors. So in the car on the way back to the Farm, Skip and I pressed her for details, anything she could remember. There wasn't much: "He just pulled me aside and said, 'There's something I think you should know. My last assignment for the Bureau was investigating you guys.'"

We would never have known if he hadn't decided to tell Carol. John Glennon was the guy with the camera in the driveway Annie had seen through the kitchen window our first summer in Lanesville. He had spied on us and we had voted for him.

11
Not a
Crook,
a Spy

The cop kicked him two, maybe three times, expecting him to move on. But the wino didn't break from his slouch in the tenement doorway on New York's Lower East Side. He cursed the cop in a resonant baritone. The cop hesitated. "You on the job?" he asked.

"Yeah," grumbled the wino.

Glennon rocked back in his chair and laughed as he recalled that encounter in 1972, while he was still a special agent for the FBI. He didn't bother to mention why he'd gone under cover as a derelict, and because the retelling happened almost a decade after the incident, I let the omission slide. Glennon had long since left the FBI, had finished with town politics after two terms as supervisor, and had gone on to become principal of Hunter-Tannersville High School. Videofreex no longer lived in Lanesville, but I had sought him out ostensibly as the subject of my first piece as a free-lance reporter for the weekly *Woodstock Times*. The real reason was because I wanted his help.

We'd never told his story on Lanesville TV. We hadn't wanted to draw attention to the fact that the FBI snooped around our house. We accepted as a matter of faith that all radical organizations, including ours, were the target of government spying, and we certainly appreciated the irony of knowing–and having helped elect–the guy who spied on us. We told our colleagues about Glennon's visit to our house, testing in each case to see whether our willingness to coexist in the same community with him made us politically suspect. That same wariness compounded when it came to our Lanesville neighbors. We knew instinctively that whether we shrugged it off or condemned it, any public acknowledgment that the FBI had spied on Maple Tree Farm would not have improved our standing in the community.

Glennon's stint "on the job" lasted from 1971 to 1973. He joined the Bureau as a special agent when J. Edgar Hoover still reigned as director. And though Hoover died in May of 1972, he and his agents remained unimpeachable American heroes until the backwash of Watergate revealed the director's depravity, the corruption at the top of the Bureau and the injustices inflicted by its agents on innocent people in the name of national security. We figured that if our neighbors knew the FBI had us under investigation, they would automatically assume we must have done something wrong. Just what the Bureau would want them to think. This analysis overlooked the point that at least some of our neighbors already knew the FBI had shadowed us, because somebody in Lanesville supplied the special agents the superficial information the agents sought.

The wino get-up was only one of Glennon's disguises. He'd also posed as an assessor's assistant in the nearby Ulster County community of Boiceville, in an operation he said was part of the Bureau's pursuit of radical fugitives the FBI believed were hiding in this area of the Catskills.

He told me this casually as we sat in his tiny principal's office. Prints of Norman Rockwell and Currier and Ives lined one wall, and secretaries popped in at awkward moments, bringing unwelcome administrative details to his attention, breaking the train of his memories. But Glennon warmed to the subject quickly, telling me that from mid-'72 until the following August, he was part of a special intelligence unit of the FBI's New York office known as Squad 47. The squad had located a safe house in Boiceville he believed was used by the Weather Underground to hide fugitives on their way out of the country. Glennon approached the house on Traver Hollow Road in the guise of an assessor to gather information about the layout of the place. In the process, he discovered a trail leading over Cross Mountain into Woodland Valley near Phoenicia.

He had suspicions about Woodland Valley for two reasons. A helicopter reportedly made regular nighttime landings somewhere in the valley, and the proximity of these mysterious flights to the alleged safe house concerned him. The second reason he was curious about Woodland Valley had to do with a Bureau report that alluded to "a lot of communal activity" in the area. The mere existence of this activity provided justification enough for him to investigate people throughout the area living communally, or collectively or in whatever other fashion the Bureau chose at that time to define as deviant or subversive. In shades of the IRS, it cannot have escaped his attention, as well, that by observing some of the more private aspects of this communal activity, he might well produce a field report full of titillating details, guaranteeing it would go right to the top, to Hoover, the paramount voyeur. But this possibility didn't enter our conversation that afternoon in his office. I didn't want to distract him.

Glennon's first task involved recording the license plates of cars and other vehicles parked at the communes to see whether anyone traveled between there and the safe house. Never mind that this form of surveillance presumed the people who lived communally had any interest in–let alone interest in helping–the Weather Underground.

Rainbow Farm lay at the end of Muddy Brook Road nestled in a sloping meadow between Panther and Garfield mountains near the mouth of Woodland Valley. The people who lived there had far less interest than we did in politics. Most of them were building their own homes and living as simple lives as they could devise for themselves. They had no radical ideology and hadn't played a role in the antiwar or civil rights movements. Their experience with spying was restricted to the discovery one summer day of a town highway crew hiding in the bushes, sharing binoculars as the workers watched the Rainbow Farmers gardening in the nude. But no matter who you were or what your political sentiments, if you visited any of the communes in the region during the early 1970s, there's a good chance John Glennon took a snapshot of your license plate and filed it with the FBI, where it remains today.

I liked Glennon. He has a friendly manner, an unaffected forcefulness that adds a measure of credibility, not always warranted, to everything he says. In all the time I knew him, he never displayed the pinched, taciturn arrogance with which most big-time cops treat outsiders, especially outsiders they've investigated. When he told Carol at Town Hall in Tannersville that he had investigated Maple Tree Farm, it seemed a genuine gesture of friendship, a way of building trust. He told her he'd been assigned to photograph our license plates because, "We found your name on an envelope in the Manhattan office of a radical organization." He didn't volunteer more information and the revelation startled us so much, nobody had the presence of mind to probe further, then or later.

I'd always associated his photographic visit with our arrival in Lanesville in the summer of 1971, but by his reckoning, he hadn't been in the area then, nor had he arrived in Lanesville until a couple of years later. By that time at least one branch of government law enforcement already had our names and knew our whereabouts. When Skip, Nancy, Bart and Chuck went to Miami to cover the Democratic and Republican national conventions for TVTV in 1972, they voluntarily submitted their names, addresses and social security numbers to the Secret Service as part of the floor pass clearance procedure. If FBI agents had an interest in collecting data on the residents of Maple Tree Farm, they could have started with a call to their fellow agents at the Treasury Department, though the records do not indicate anyone ever bothered to make that call.

We all knew about the photographer in the business suit Annie saw from the kitchen window and the guy in the diarrhea-green truck tapping our phone, who left when Davidson began snapping pictures of him. No one got exercised about these incidents. At worst, they constituted an annoyance; at best, they confirmed our status as important players in the antiwar movement. Had the FCC shown up, we would have had real problems. With any other government heavies, well, we had work to do and nothing of a political nature to hide.

∞∞∞∞∞∞∞

Carol and I had moved out of Lanesville in 1975 and did not see Glennon again until we him met by chance on Tinker Street in Woodstock on a sunny Sunday late in July 1977. He greeted us as friendly as ever, eager to talk. A federal grand jury had recently indicted two top officials with the FBI's New York City office, charging them with ordering illegal break-ins during the late 1960s and early 1970s. This sounded to me like something Glennon might have had his hand in, and I brought up the topic as we spoke on the street. He said he was sure that if anything, he'd be liable only for civil penalties not criminal ones. But he quickly pointed out that he'd worked at the very center of the "Weather-fuge" investigation, which was mentioned in the indictments of the top New York agents. "I was a cat burglar with a license to kill," he said with cheery pride. That was pure John Glennon–springing information on you in a way that confounds an adequate reply.

Weather-fuge was one of a series of COINTELPRO operations dating back over a decade, which the FBI mounted against groups and individuals it believed represented a threat to the security of the nation. No one outside the Bureau other than a few high government officials had ever heard of COINTELPRO until 1971. But on the night of March 8 that year, while Joe Frazier outslugged Muhammad Ali in Ali's comeback after years of forced idleness because of his draft resistance (we were busy that night projecting the fight at a theater in the Bronx), a group of activists calling themselves the Citizens' Commission to Investigate the FBI broke into the Bureau's office in Media, Pennsylvania, a Philadelphia suburb. The burglars stole hundreds of files and then released them to the press. The documents revealed a widespread, officially sanctioned program of illegal domestic spying and related unlawful activities–safecracking, wiretaps, mail openings, IRS investigations, and more sinister acts of provocation–all of them conducted by FBI agents and their informants.

COINTELPRO stood for Counter Intelligence Program, and Hoover directed it not at suspected foreign agents but at citizens of the United States and lawful organizations to which they belonged. The COINTELPRO revelations briefly made the news, but did so well before Watergate became a household word, and the mass media displayed no enthusiasm for pursuing a story of potentially illegal activity by the chief law enforcement agency of the country. Despite a growing national wariness of outlandish military optimism on the progress of a war we were losing, the general public did not yet understand the extent to which we were being force-fed propaganda, or the lengths to which the government had subverted the democratic process. The records stolen from the Media files led a few journalists, politicians and civil libertarians to express genuine shock that the federal government would spy so indiscriminately on its citizens. But by and large, the mainstream press treated the break-in as antiwar vandalism and ignored the implications of the illegal activities the files uncovered. The internal government reaction, however, was quite different.

Three months after the COINTELPRO files got out, *The New York Times* won a Supreme Court decision that allowed it to publish the Pentagon Papers. The Pentagon Papers was the name applied by the news to a secret Department of Defense report, "History of U.S. Decision-Making Process on Vietnam Policy," leaked to the *Times* by Daniel Ellsberg. Here, too, the contemporary controversy about the Pentagon Papers case did not focus so much on what it revealed–years of specious body counts and worthless promises of victory over the North Vietnamese had already undermined national illusions about the righteousness of our involvement in the war. Instead, lawyers and politicians argued over whether the *Times* should be allowed to share classified government documents with the public. The attempt by Nixon's attorney general, John Mitchell, to bully the *Times* into killing the Pentagon Papers story failed, and the high court's decision overruling the administration's demand for secrecy fed Nixon's mania for using any means necessary to deal with his domestic "enemies."

The release of the Pentagon Papers, following on the heels of the break-in at Media looked to Nixon and his advisors like an overall conspiracy linked to sporadic incidents of urban and campus unrest around the country. As if to underscore this theory, the protest movement now had an identifiably violent splinter group. The Weather Underground had accomplished the unthinkable: a wave of politically motivated terrorist bombings around the United States, including a powerful explosive device detonated on the Senate side of the Capitol. The government launched a frenzy of internal security actions, with the White House creating its own secret operatives because Nixon didn't trust the FBI. And the FBI, unchastened by the COINTELPRO revelations, grew more determined than ever to break up what was left of the antiwar movement. John Glennon stepped into the middle of this mindset of security paranoia when he joined the Bureau.

COINTELPRO didn't spring from the antiwar protests of the late 1960s and early '70s, nor did it arise in response to the uprisings and riots that gripped American cities in the mid-to-late 1960s, although it reached its zenith during those periods. COINTELPRO began officially in 1956, when Hoover won the approval of President Dwight Eisenhower and his vice president, Richard Nixon, to spy on and disrupt the activities of communist organizations in the U.S. Hoover never showed much interest in pursuing organized crime; subversives, as he determined them, were his primary target. He'd had his first taste of raw power as a leader in the notorious Palmer Raids following the First World War, aimed at people the government determined were politically undesirable. After he was placed in charge of the newly created FBI, Hoover gathered publicity for the agency with its apprehension of a few, big name gangsters and by fostering movie and radio portrayals of special agents as crime-fighting supermen, incapable of making a mistake unless it involved getting killed in the line of duty.

John Glennon didn't precisely match the publicized career path of the special agent. He came from Tannersville, served in the Marines and worked as a guard at the Coxsackie prison for juvenile offenders. He hadn't gone to law school. His degree was in art history from the State University College at New Paltz. At the time he joined the Bureau in 1971, he was teaching and had a gun business on the side in the town of Hunter. At first he was assigned to criminal work, and he'd had his picture in the papers hauling a robber out of a bank in the Midwest. He claimed that was the most publicity he'd had until he appeared on Lanesville TV as town supervisor.

∞∞∞∞∞∞∞

I lost track of Glennon after that brief meeting in Woodstock, but was not surprised to see his name in *The New York Times* a few years later. In October of 1980, *Times* reporter Robert Pear ran a piece under the headline "Reporter's Notebook: A Trial That Often Seems a Flashback to the Early 1970's." Pear wrote down his impressions from the trial of former FBI officials W. Mark Felt and Edward S. Miller, who were accused of ordering illegal break-ins as part of their investigation of the Weathermen in 1972-73. This was the trial of the indicted agents Glennon and I had discussed briefly three years earlier. Glennon was called as a witness and, from the perspective of the government that once employed him, not a very cooperative one. According to Pear's account, one exchange between Glennon and chief prosecutor John W. Neilds Jr. went like this:

> "Did there come a time when you did a bag job on the home of Jennifer Dohrn?" Mr. Neilds asked.
> "I object to the adjective 'black bag job,'" Mr. Glennon replied. There followed this exchange:
> Q. What is a black bag job?
> A. I remember reading about a black bag job. Something about a burglar running around with a little bag of tools, doing something he's not supposed to do.
> Q. Did you do a black bag job on Jennifer Dohrn's home?
> A. No, I did a surreptitious entry.

When I interviewed him in 1980, not long after that story appeared, Glennon said he finally conceded the semantic point that a black bag job was a surreptitious entry, but he insisted what he had done was neither illegal nor unethical. With a flash of angry recollection, he said to me, "God damn it, I wasn't a crook. I was a spy."

He understood the distinction between crooks and spies in clear terms. He likened the period he served as a special agent to a war, admitting that his side was losing most of the time. Squad 47 did not achieve conspicuous success. Time, not the FBI, brought most of the surviving members of the Weather Underground into the open. Not long after I spoke to Glennon, Bernardine Dohrn, Jennifer's sister, and her companion, William Ayers, surrendered to the authorities. They had been living on the Upper West Side of Manhattan, and they may well have wanted to cut a deal during the last days of the Carter Administration rather than face the possibility of a prosecutor and judge appointed by incoming President Ronald Reagan.

Bernardine Dohrn had jumped bail after the "Days of Rage" in Chicago, which we had caught only part of when we went to Chicago for Don West in 1969. She had last been seen by anyone willing to identify her in an independent film on the Weather Underground made in 1975. Even with that kind of exposure, the FBI had no luck locating her or other Weather Underground fugitives.

It was not from lack of trying. Glennon acknowledged the difficulties he encountered in his hunt for the fugitives. Some of the information he gathered was tainted by the methods employed to get it and couldn't be used in court. "In intelligence-gathering you're not doing it for the courts. You're doing it to locate someone. Of course, there is always that gray area."

I thought for a moment his reference to a gray area might indicate some small sign of remorse or, at a minimum, a certain inner conflict. I had trouble imagining how anyone could do what he did and not have second thoughts. He had told me about a woman in the Ulster County hamlet of Mount Tremper who had a baby with a birth defect. He said she was the companion of a radical fugitive he identified as Pun Plamondon, a member of the radical group known as the White Panthers wanted in connection with the bombing of a CIA facility. The woman had no money and depended on public assistance. Glennon said he and another agent threatened to end her welfare benefits and arrange to have the baby taken away from her if she didn't cooperate and tell them where they could find their prey. But no, he said, no second thoughts. "It didn't bother me in the least."

What was the point of all these investigations, the break-ins, the threats and the spying?

"We kept so much pressure on 'em we made 'em paranoid."

That rang true, and it didn't apply only to the fugitives. A legacy of paranoia lingers even today in the lives of those who became the targets of investigations like his, surfacing among people I know in half-hearted jokes about whose phone tap is malfunctioning when there's static on the phone line. It comes to life in a palpable way every time I come upon the information the FBI collected on our life at Lanesville, the stuff they wouldn't release and the two memos they would. I passed both the memos across the cluttered desk for Glennon to examine.

This was why I had wanted to see him. One of the memos listed me and all the other Videofreex at Maple Tree Farm. On both, the thick black line of the censor's marking pen obscured the names of agents and informers and of information the nature of which I could only guess. I hadn't thought to go looking for my FBI files until I mentioned the Glennon story to Woodstock attorney Alan Sussman on the train to New York City earlier that year. Sussman, an expert in constitutional rights, offered to seek the files for free. I had no intention of suing Glennon or anyone else, and Sussman didn't suggest it. I simply wanted Glennon to help me understand why he felt it necessary to spy on us and what his spying had produced.

He squinted at the memos with a practiced eye, seeing meaning in notations indecipherable to me. The first memo was dated March 16, 1973, and I assumed he had filed it. It consisted of three pages, two of which were a copy of a form letter we had mailed to all the post office boxes in Lanesville a month earlier. It introduced us as members of Media Bus and attempted to explain how we made our living. That winter had been especially hard. The Council had made further cut-backs in our funding and compounded our financial plight by delaying payment of promised money. We scrambled to keep the group together even as individuals were making plans to go their own ways. It got too difficult to get a show together each week, so we temporarily went off the air. Defensively, our letter read, "As the community became more isolated from us, we became less able to provide good programming and less in touch with what was happening in Lanesville." We said we wanted to resume broadcasting, and we wanted to know how we could do that without interfering with other reception. A questionnaire had been appended to the letter, although it did not make it to the FBI files, or, possibly, was withheld for "security reasons." Each of our names was listed at the end of the letter. It was about as thorough a synopsis of activities as any spy agency could want. But it didn't satisfy the Bureau.

The FBI cover memo dutifully copies each of our names, with the exception of Carol's. She had decided to place her name on the form letter as Carol Vontobel Teasdale. Even though she seldom used my name, it served as a nod toward social convention that dismissed any gossip the two of us lived in sin. The FBI memo carefully changed her name to "Carol Teasdale nee Vontobel." The rest of the memo summarized reports on us from one or possibly two informants, one of whom "commented these subjects have not caused any problems in the area and appear to be polite and well educated." The first informant issued a report to the FBI on the date of my 25th birthday, noting Media Bus consisted of "9 persons, all white, hippie-type life style." The awkward language may well reflect the haste of bureaucrat-spies, but it could, as well, indicate the reported difficulty the Bureau had in identifying exactly who comprised the enemies of the state.

∞∞∞∞∞∞

The delicious part of the FBI's memo, and the one that still strikes me as mysterious, is the assertion that we polite devotees of the hippie-type life style "produce local TV shows which are frequently seen in the area on a local TV channel." On the face of it, the statement is certainly correct. But our letter talks about our transmitter and asks the readers to suggest the best channel for us to broadcast on. The agent had got the names and the reports of informers, and that satisfied his requirements. No one at the FBI seems to have figured that we ran the "local TV channel" as our one and only truly subversive–and illegal–activity at Maple Tree Farm. I was looking forward to a laugh at Glennon's expense on that point.

"Well," he said, still gazing at the photocopied pages, "this isn't my memo." I did a double take. He went on, "This was filed by the resident agent in Kingston. We followed up our own leads and then just passed along the information. But I know I never saw this letter." He shook his head apologetically.

I swallowed my unlicensed transmitter story and conjured in my mind the image by which a file is passed from one agent to others in other departments, in other agencies. I saw it grow along the way from a license plate number to the report of a neighbor to become something like the second memo, the only other information the FBI would agree to release. It was dated November 11, 1977, more than four years after the first one.

The censor used a heavier hand with this second one, obliterating all but a half dozen lines in a document that ran for more than a page. I had asked Alan to file an administrative appeal of the deletions, and we were successful to the extent that, like the skin of an onion, another few inches of marker fell away. The new version revealed the inanity of miscellaneous items pertaining to Maple Tree Farm. Its source was most probably a guest of ours, in which case our spy was someone we fed and put up for the night. The contact had been made a week earlier than the date of the memo and the report was filed under the FBI heading: "MAPLETREE FARMS, LANESVILLE, NEW YORK, MISC–INFORMATION CONCERNING." It includes the discovery that some persons whose names are crossed out "and a few other individuals are residing in a 'commune type' atmosphere at the Mapletree Farms." The informer dutifully reported four license plates, one of which was a Media Bus car. The other plate numbers are crossed out. The unnamed spy gratuitously "advised he was furnishing this information for any value it might have in an investigation."

The report concludes, "Albany indices are negative re the aforementioned names and it is recommended that this memorandum be filed for information purposes." Just in case.

Glennon puzzled over this memo for several minutes. He pointed out the obvious, that it had been filed long after he had left the Bureau and even after the Bureau had withdrawn its resident agent in Kingston.

Carol and I lived in Phoenicia by the time this spy had dropped by, and had for two years. David, Davidson and Annie were gone, too. Lanesville TV limped along sporadically.

I told Glennon I knew he hadn't been involved in this report, but I would appreciate his observations about its contents. He pointed to a handwritten number at the lower right corner of the first page. He said he could not recall exactly what the first few digits stood for, but he thought it designated an undercover source. "I think you had an informer in your group."

I asked him for the memo so I could re-read it. It never occurred to me it might have originated with one of us. I stopped myself. I didn't need to see it. I had memorized its incomplete sentences, ones like, " ... and are supposedly associated as writers with ... ." I knew without looking at it again that I could not have spent so many years living and working with these people, my best friends, only to have them describe me or each other like that. Glennon had misinterpreted the words, the type of misinterpretation that keeps the pressure on and makes people paranoid.

By his own account and the available evidence John Glennon had an exemplary, if brief, career as an FBI agent. He received two personal citations from Hoover for his work on criminal cases. He had been a good town supervisor, intent on solving problems; and I think he was probably a good high school principal, too, not overly concerned with petty details and more in touch with the real world than many of his colleagues. He said he was bitter about what happened to some of his superiors at the FBI. The investigations and the criminal prosecutions that followed hurt everyone who served in Squad 47. But I didn't sense bitterness drifting through our conversation as much as a longing for the life of a special agent. He'd rather be back with the Bureau, he said.

Before I left, he made a copy of the *Times* article quoting his testimony. Then he walked me down to the parking lot. He had to leave for a basketball game, Hunter-Tannersville against nearby Gilboa. He said if I needed any further information I should call. He was serious about the offer.

Alan told me a few days later he had little hope of dislodging more information from government files. The law protects the privacy of informers. Occasionally, though, I hold my copies of the FBI memos up to a bright light just to see if I can make out anything beneath the slash of the censor.

12
Buckaroo
Bart

Our first regular Lanesville TV broadcasts began promptly at 7 p.m. on Sunday and Wednesday nights and lasted for as long as we felt we had something to show. Some broadcasts rolled along for an hour or more. Others struggled through twenty minutes before we pulled the plug, mercifully sparing neighbors, visitors and ourselves the pain of stretching weak material or no material at all to fill some arbitrary time slot. It wasn't as if other programs were scheduled after ours. Lanesville TV, Channel 3, went dark when we were done, freeing us from the tyranny of the clock that rules other broadcast media.

Before long we dropped the Wednesday shows as too disruptive to our work schedule–the work we had to do to make a living. The disappearance of Wednesday night Lanesville TV induced no reaction at all from our viewers, no outraged phone calls, no sighs of relief, either.

In the meantime, Bart had come up with the idea for *The Buckaroo Bart Show*, a Saturday morning children's program. It starred Bart in the title role and included John Benjamin, a particularly outgoing local ten-year-old, in the role of Sheriff John. And as no TV show, especially one that purports to be about cowboys, would be complete without a villain, Bart drafted Howard Raab, a skilled carpenter and stained glass artist who happened to live in the cabin behind Maple Tree Farm. Bart assigned him the role of a comic antagonist by the name of Horrible Howard. Horrible, as his name became shortened both on and off the set, had a bushy beard that protruded from his chin, and a heavy brow he could knit into a caterpillar underlining his forehead. He affected a bowlegged walk to accompany his thin grin and slightly nasal voice, all of which marked him as an inept bad guy, not an evil one.

The first *Buckaroo Bart Show* was a short, improvisational morality play called "Don't Throw Yer Cans in the Road." Nancy shot it on Neal Road, the short strip of pavement running from the highway to the far end of our driveway. The scene opens with the stars sauntering down the road toward the camera, the sheriff on one side of Horrible and Bart on the other. Horrible finishes his softdrink while he's walking and flips the empty can over his shoulder. Our two heroes are so shocked by his scofflaw behavior they stop dead in their tracks, while Horrible keeps loping along. With a little prompting from Bart, the sheriff, a pudgy guy about as tall as Horrible's solar plexus, grabs the villain by the arm and escorts him back to his can, where he and Bart lecture Horrible about the evils of littering. Both the medium and the message were new to John, who nonetheless handled them with the aplomb and self-assurance of a pro.

The Buckaroo Bart Show slowly grew into an elaborate production involving science fiction plots and complicated location shooting. J.P. was pressed into service for a series of episodes in which he invents a "transporter" that Horrible inadvertently activates, sending Bart and the sheriff to the moon: "The moon, Horrible?!?"

"Yeah," says Horrible as deadpan as ever, "the Moon."

There was also a Halloween show, "Captured by Ghouls", in which Horrible, Bart and the sheriff are chased, Marx Brothers style, in and out of the many doors along the narrow second floor hallway at Maple Tree Farm. The ghouls were visitors to the Farm that weekend–other video people given masks and minimal direction before being turned loose in front of the camera to ham it up.

The cast occasionally rehearsed the unscripted dialog so they could grasp the broad outlines of what passed for a plot. And for the "Lost on the Moon" series, there were cliffhanger endings, with our heroes left in some perilous predicament that seemed hopeless. Tune in next week.

In its early days, commercial TV dabbled occasionally in improvisation, more often than not when some comedian like Milton Berle or Sid Caesar blew a line or stepped out of character accidentally on purpose, giggling, squirming, struggling to get it together and make it to the next commercial break. By the early 1970s, the spontaneity had been wrung out of television. Comedy had moved from live to strictly film and then to videotape, with *All in the Family* the most popular show. Video did not mean live or unrehearsed, and it certainly didn't mean unpredictable. The Smothers Brothers had been as radical as CBS had dared get, and the brothers had paid the price for having taken the most innocuous of stands. Abbie had been visually censored in 1970 when he appeared on *The Merv Griffin Show* wearing an American flag shirt. CBS had ordered the screen blacked out when the camera was on Abbie, which only called more attention to his appearance. (No such blackout was enforced when cowboy star Roy Rogers, a darling of the right wing, appeared on another network in a similar flag shirt.) What passed for spontaneity on entertainment programming was limited to infrequent tapes of network-sanctioned screw-ups on out-takes of *The Carol Burnett Show*.

Part of the networks' aversion to impromptu programming had to do with their craven behavior in the face of even the possibility of political criticism by the federal government or right wing zealots. Anyone who thought the TV Blacklist had died with the withering of McCarthyism needed only consult Tom and Dick Smothers to be convinced it lived. At its core, the issue came down to greed. The networks and the companies that produced shows for them made millions on successful series, and they wanted nothing in their control left to chance. Live or improvisational programming is by its nature chancy–not a good financial risk.

The Buckaroo Bart Show had no budget, making it an improv affair by necessity rather than choice. It had the look of a funky kids' show, but the similarity went only skin deep, because the programs had to rely on unpredictability, which no scriptwriter could have imparted. That unpredictable quality extended to the scheduling of the shows, too. With none of our grant money specifically earmarked for *Buckaroo Bart,* or for any other Lanesville TV production, the kids' show project went from a regular event to an occasional program and then, as suddenly as it had arrived, it disappeared altogether.

∞∞∞∞∞∞

The weekly evening broadcasts survived primarily because we usually produced them as an afterthought. By the spring of 1972, we had polled our viewers on the air, and decided, based on the half dozen or so responses, we should switch from Sunday to Saturday night. That seemed to boost our audience, especially in the winter, when the skiers were around. In the summer, when it was still light outside at 7 p.m., we learned the same lesson the networks had years earlier: Fewer people watch TV in the summer. So we went on the air an hour later. One advantage of summer was that we could use the front porch and lawn as our studio. We even, on occasion, resorted to a rerun, although always with a live host and an open phone line.

Our weekend house guests and occasionally a few local folks would assemble in the viewing room or, in the warm weather, on the porch, to watch the show. Most would stay afterward for one of Annie's ample dinners. Later we would gravitate to the viewing room, as soon as Chuck had gone to bed, to preview a visitor's latest tape or show off one of our own. Some nights, when the spirit moved us, we made music. Skip had bought a conga drum, Davidson contributed a spinet from his parents' house and gradually we accumulated a plentiful supply of maracas, tambourines and other hand-held percussion instruments to accompany the pump organ. On those nights, the music room rocked with boisterous drumming and singing, the strains of a recognizable tune filtering through now and then. We wrote our own songs, too. I composed a tune for Carol's whimsical documentary on cows, and Skip and I labored for several nights in a row on the lyrics. Chuck walked by and earnestly suggested the line, "Cows give you milk, butter and eggs." I loved it, but it was vetoed by Skip and Carol.

Factions would split off. David and Davidson would invite selected guests to retreat to the third floor. Following the departure of Frances, Davidson had knocked out the wall between his room at the back of the house and the small adjacent bedroom. He'd demolished the ceiling, too. Sam scratched his head, but gave the project his blessing. He respected Davidson's ability as a carpenter and Davidson paid for the materials himself. After months of fitful activity, he stopped work on his mini-loft. The wallboard remained mostly unpainted and the windows, including a new one high up on the north side of the house, never received more than rough framing. He created a loft where none had existed, defining a garret-like space that fit his own persona as individual artist. He left the fine points to the imagination and could not be bothered putting the finishing touches on his quarters. His was the largest, most intriguing room on the upper floors, with steeply slanting walls that reached clear to the peak of the roof. In his loft, he and his invited cognoscenti would blow a joint and plot video strategy, scoffing at the uptight "Presbyterians" on the floors below and our Babbit-like flirtation with pop culture-broadcast TV. If one of us ventured uninvited down the long, dark hallway leading to the doorway of the loft, its denizens abandoned their conspiratorial chuckling and extended a warm welcome to the intruder.

David and Davidson, as the two eldest members of the group, felt their views were not given appropriate weight as we struggled to keep Videofreex afloat year to year. To me, the youngest member of the group, that just made them seem older and more out of touch with the direction I saw the group heading–toward greater involvement in Lanesville TV.

We never made a dime on programs like *Buckaroo Bart*. Not directly, anyway. The people who supplied the funding from the State Council on the Arts showed up every year or so to check out our activities as we'd described them in our grant applications. This kind of bureaucratic snooping might have offended us had Russell Connor and Lydia Sillman, his assistant and later successor as head of the TV/Media Program at the Council, not been artists themselves. Russell had been the assistant director at the Rose Art Museum at Brandeis University and was responsible for the *Vision & Television* show in early 1970, the first major museum show of video art. Annie had worked for him and that's where we met her. Russell professed to be ill at ease with the politics of the Council, especially the infighting among video groups desperate for funds. He also seemed more intent on proving he was as hip as we, sometimes at the expense of forcing us to demonstrate we had spent the state's money wisely. Lydia was a dancer, and the troupe she was part of performed one summer in a Woodstock stream. A photo of the event captures naked dancers cavorting in a waterfall just above Davidson, who is hunched over a video camera on a tripod, wearing a t-shirt, jeans and brown rubber hip boots.

Council video officials knew all about Lanesville TV, knew it had no license and didn't seem to care as long as we didn't embarrass the agency by getting into trouble. In those waning years of a wartime boom, even after Rockefeller quit as governor and ascended to the vice presidency, and even as the Arab oil embargo took hold and the economy began to stall, the money continued to flow from the Council and, later, from the National Endowment for the Arts. On our grants, we identified ourselves not as Videofreex but as the members of Media Bus, only to drop that name, too, for all but strictly legal matters, substituting instead the institution of The Media Center at Lanesville, of which Lanesville TV was but a project. The grant amounts diminished over the years, and we had to spell out with greater detail the uses to which we put the funds.

Yet we always had just enough left over to allow us to keep making shows like *Buckaroo Bart*, and other loonier pieces for which the cowboy show served as a prototype.

∞∞∞∞∞∞

Where network shows live and die by the ratings, as often as not, we measured the success of shows like *Buckaroo Bart* by the amusement we derived from them. If a plot was fun, Bart pursued it, with Nancy sharing the directorial chores and shooting each episode. It might as well have been live for all the editing we did. MTV and other cable networks rediscovered the energy of this approach in the 1990s, with shows that record the continuing saga of the lives of young people. The glitches and crises–or enough of them to make the MTV shows look unrehearsed–are left in. We retained our glitches because we had no other option. Often, *Buckaroo Bart* shot one take for each sequence, not unlike two-reeler cowboy and adventure serials Hollywood cranked out in the 1930s and '40s. The editing process, now as then, remains time-consuming and technically formidable, and while MTV has learned how to produce low-budget shows, we made do with no-budget shows.

Our free-for-all approach to *Buckaroo Bart* prevailed as well on the regular evening shows. If we wanted to try a new location or a new camera set-up on the Saturday evening broadcasts, whoever had the idea was free to try it out provided he or she could persuade others in the group to help.

The original concept for the evening shows had been to have the on-camera host at the switcher in the control room, much like a disk jockey at a radio station. But a TV program requires more complex operations than a radio broadcast, and small distractions could create havoc. I found myself more than once pausing on-air, mid-sentence with my finger poised to press some crucial button, unable to recall which button I was supposed to hit. I went for the wrong one more times that I care to remember.

The conventional model for TV show production relies on a rigid hierarchy. A producer sits in an isolated control room looking over the shoulder of a director who tells a technical director to switch between cameras located out on the studio floor. The camera operators must train their lenses on precisely what the producer wants. We had long ago rejected this system as creatively stifling. But our initial experiences with Lanesville TV shows originating from the control room led me to believe the idea of a separate technical space might have some merit after all. The isolation allows the people handling the mechanical aspects of the show to concentrate on their jobs rather than dividing their attention between the technical work and the need to remember not to stare off into space or drool on camera. Likewise, the people in front of the camera can concentrate on connecting with the audience without the distractions of the panicked dialog that kept our shows on the air: "Why is the picture so crappy?" "Oh no, Sam says there's a buzz in the audio!"

We compromised on this point. Rather than surrendering entirely to tradition, we set up the control room with all the VTRs and monitors and other production paraphernalia in the front dining room. We designated the adjacent middle dining room as the studio. (Occasionally we moved the furniture and reversed this configuration.) The rooms were separated by sliding doors with French windows, so in theory we could isolate them from each other. In practice, however, that seldom happened. The shows went on the air and the people in front of the camera called out questions and instructions to the folks in the control room just as you'd speak to anyone in the next room. And the folks in the control room would answer. Crises in the control room–and we had plenty of them, as VTRs gobbled up tapes, mics went dead or the transmitter begged for someone to kick it–became part of the show. We thought this demystified the process of television, a political goal of ours. To produce a show without the hierarchy amounted to a revolutionary act directed at the oppressive mediocrity of conventional TV. More important, it would point the way to a new form of television. Perhaps if we'd made money at it, our approach might have had a more immediate and noticeable impact.

I'm no longer so sure people want their entertainment demystified. And in any case, the ubiquitous, portable VCR has put video quite literally in the hands of the masses. CNN and some of the TV networks have made a big deal of showing off their control rooms as a visual element of their news broadcasts. But these images, which portray hushed, dimly lit and thoroughly isolated control rooms, just underscore how strong the barrier between technology and "talent" has remained in most studios. The chitchat between host and control room made our broadcasts homier, a characteristic parodied now by the network shows like *Saturday Night* as the mark of cable TV ineptitude. When corporate TV figures out how to make a homey style fabulously profitable, every show will adopt it.

Things got a little hairy whenever we shot shows in the kitchen, a logistical problem because we had no functioning intercom. The ancient intercom we inherited from the CBS project, complete with clunky telephone operator headsets that got tangled in our long hair, hurt our ears and invariably drooped to our necks at critical moments. It never worked right. It produced inaudible commands or else blasted in our ears at a deafening volume easily picked up by the microphones on the set. So we scrapped the whole intercom concept and agreed in advance what each camera operator would be responsible for shooting, leaving the rest to the member of the group who had the job of switching between cameras. This method of putting together a crew also undermined the highfalutin titles that go along with jobs in conventional TV production. None of us would have dared call ourselves the director. Someone would call out, "I wanna do the switcher tonight," and he or she got the job.

This approach fit with our free-form attitudes about television and satisfied our individual egos in that no one of us ever got to tell the others what to do. Our version of a multicamera TV show amounted to a kind of umbilical anarchy. Sometimes it worked. Sometimes it didn't. Either way, we and our viewers always anticipated an element of surprise.

The process of selecting tapes and a host for the show each week developed along equally unstructured lines. A lot depended on what had happened that week. When Skip's grandmother Lutz, a small, white-haired woman with a wry smile, came for a visit and sat in as a guest, it was a foregone conclusion that Skip would be the host that week. Likewise, when my mother stopped by, I did the honors. No one fought over the right to host a show. We always had room and time for one more. Mostly, it fell to the person who started thinking about the show first to suggest a host.

Davidson probably hosted the fewest shows, although he did make a memorable appearance as a trout, with his face peering through the head of a fish in profile drawn on a large sheet of paper. Chomping on a cigar, he gleefully taunts fishermen and offers a fish's-eye-view of human flycasting. That bravura performance aside, he, of all of us, felt the least comfortable in these group efforts that had so little to do with art. He had set himself on a path of exploration into the aesthetic possibilities of the medium, and Lanesville TV was a distraction. He never intended his multi-monitor pieces for display on the crass medium of TV. They had a sculptural quality and a time sense outside the normal confines of entertainment. When he agreed to be the host, his pursed lips and the slightly judgmental tilt of his head lent him a not altogether unintended air of superiority. After Curtis, his was the first disaffection within the core group.

Chuck, too, hosted only a few shows. He preferred to work off camera in the control room, fussing over the signal and troubleshooting the cascade of minor disasters that afflicted most broadcasts. He did agree to appear on camera, but like me he was stiff. He had no gift for small talk, either, and would frequently rely on challenges to viewers—Hey, Sam, where are you tonight?—if the phone hadn't rung.

Ann was a reluctant host, but when we introduced *Lanesville Country Kitchen*, inviting people from the area to cook their favorite dishes on the air, her involvement increased and her easy laugh and smooth, pretty features would have made her a natural had she wanted to do more work in front of the camera instead of behind it. Ann had made her place in the group by being the least vocal, by offering her skills tentatively and withdrawing whenever it seemed she might be in conflict with anyone else. Many of us suspected she had hidden talents and I took it on myself to test this theory. I pressured her to edit a tape with me, teaching her the basic skills as we went. During that edit, I repeatedly asked her advice to make the calls on where to cut the pieces, while I handled the technology. Slowly and with considerable hesitation she began to get the hang of it. I could see she had the aptitude, but neither I nor anyone else could persuade her to take on a video project of her own for as long as she lived at Maple Tree Farm.

Bart hosted the first *Lanesville Country Kitchen* segment. The guest cook was a woman named Fran Groenwald, who lived down the road and who volunteered to prepare her cinnamon swirl coffee cake on the air one evening in the fall of 1972. Bart donned a huge, floppy velvet beret, a frilly shirt and a vest, and affected the manner of a foppish aesthete with a French-Yiddish accent. As the show opens, he strides through the kitchen door and plops himself on the island. He stares at the camera wide-eyed, his pupils noticeably dilated despite the bright lights. "Tonight we're going to tickle the parrots—uh, palates—of the people of Lanesville." From off camera you can hear shrieks of laughter. Bart is unfazed. He fumbles through the introduction of Fran and comes to her recipe. Dumbstruck for a moment, a goofy grin breaks out on his face. He hasn't the faintest recollection of what Fran plans to cook. He ad libs. "Coconut creme …."

"No!" cries Nancy in mock horror, the best indication for viewers we used no cue cards. All she can do is prompt him. Why whisper? Nobody would believe he knew what he was talking about now, so word by word, a chorus of off-camera voices walks him through it.

Maybe some prima donnas would have stalked out on a show as discombobulated as this one, but not Fran. She had moved to Lanesville from Long Island with her son and his family. She loved to laugh, loved the energy of the farm, a working class grandma in the midst of a bunch of whacked-out hippies operating a TV station of questionable legitimacy. Judging from the calls that night, the audience found her segment of the show an unqualified success, with neighbors phoning to ask for the recipe. Eva Quick in Chichester, who people in Lanesville clucked about as a busybody even as they were reciting every word and then some of what she'd told them, called in by prior arrangement to inform Fran she had won a local pool on the World Series. How could anyone switching know they hadn't found the real Edith Bunker in candid conversation with a friend?

Fran did one other show, but then her son, who didn't approve of us, forbade her to return to Maple Tree Farm. He said she couldn't watch the show, either. He didn't want her seen with us, and we lost touch with her.

Ann hosted a *Lanesville Country Kitchen* a few weeks later with a friend of ours named Maria Voorhes. Maria and her husband, Earl, lived in Chichester. She had been a star of TV and the theater in Brazil and had met Earl, a struggling actor, in New York City. They had decided to give up city life and move to the country, where Earl got a job at the woodworking factory in Chichester. When the factory closed, he worked at various tasks including construction and selling rugs in Kingston. What they loved best was to party, and soon after we met them they became frequent and welcome visitors at our Saturday night shows and dinners.

Maria insisted on starting her appearance with what she called "a leettle inspeer-Rayshun." Holding up a glass of wine, she advised the audience, "Some peoples get their inspeer-Rayshun from wah-ter and some from bee-yer. I get mine from why-in," and she chugged some more from the glass. Maria could drink anybody but Chuck under the table, and the more she drank the jollier and more passionately involved in her cooking she became. She had a round, bronze face with small eyes that sparkled. Her voice had a Carioca lilt that made her enthusiasm infectious.

In the midst of Maria's preparations, Cy Griffin called from the Bureau of Indian Affairs in Washington, D.C. For the moment, our phone mic wasn't working, and Ann had to repeat Cy's description of the occupation of that building by native Americans and a handful of their white friends. Cy pressed us to join him in Washington. He needed tape and equipment. The middle of a show was hardly the place to carry on this debate, and a long line of Indians behind Cy were impatient to use the phone. Bart and Skip decided later that evening to go, and drove all night to get there. A few days later, they returned with tapes from the inside of the demonstration looking out, of an oppressed people suddenly finding themselves in charge of the headquarters of their oppressors, of acts of courage and restraint and of fury expressed through vandalism. The viewers of Lanesville TV that night saw none of this. We had no way to relay those images. They heard only Annie's demure translation of Cy's report, and she knew as well as the rest of us how deadly boring it came across. Meanwhile, Maria was drinking plenty of why-in from her refilled glass.

No sooner had Ann wrapped up her conversation with Cy than Skip got a call from an old college chum. One of the drawbacks of not screening calls, especially when you have only one phone line, is that everyone who calls automatically ends up on the air–whether they want to or not. By this time, Chuck had fixed the phone line, and while Maria sips, Skip and Lennie chat long enough for Skip to suggest he return his friend's call some other time.

Skip now has the floor, and he decides to deliver an impromptu lecture on why no one in Lanesville and Chichester should buy iceberg lettuce. Members of the United Farm Workers union were on strike in California and were urging the public to boycott the crop as a way of putting pressure on the growers to reach an agreement. Suddenly deciding that he should make an attempt to appear objective, Skip adds, "I don't know if it's politically sound."

"We don't even know what poverty is," Maria blurts out from the other side of the room. She sounds a little surly. And with that she rambles into a polemic on society's ills. Her Brazilian dish simmers on the stove, all but forgotten. The phone rings. Sam on the line. "Every'ting is good!" says Sam, leaving it open to interpretation whether he means to include the political sentiments or just the reception.

In showbiz, timing is everything. Sam's is perfection. A derailed Maria takes another sip and resumes preparing her meal. Then she pauses, looks earnestly at the camera and says, "Don't ever cook angry. If you cook angry, just cook a hot dog." And with a wave of her hand, we're back to cooking and eating and the regular Saturday night party.

∞∞∞∞∞∞

Of all the Videofreex who hosted the shows, only Nancy and Carol developed followings. You could see immediately the reasons why. Nancy had trained as an actress, done radio and TV commercials, sung professionally and performed on radio and the stage. The lens and lights, the scanned image itself, can do funny things with beauty, magnifying imperfections into defects. But Nancy, as translated across the screen, became, if anything, more attractive. She always smiled and never lost her cool, a professional's instinct. People loved it when she did the show because she added glamour to the station.

Somewhere along the way before we met she'd opted out of the showbiz competition, the cattle call auditions, the agents running her life. She'd met Bart after his Puerto Rican sojourn, when he tried to make it as a record producer and she was part of a group with a contract to make a record. He told her funny stories, made her laugh. But she ditched him for a guy named Alan Cain, who took her across country in his VW bus and then dropped her in L.A. Alan Cain showed up at the farm once. Nancy stormed into the kitchen, badminton racquet still in her hand. "He's out there and he has to leave immediately!" she fumed. She'd already given the same instructions to someone else. He split and never came back, leaving her only his name. Bart looked relieved.

Carol's sweet face and soft voice in no way prepared you for her unadorned straightforwardness, on the air or off. She spoke her mind in a guileless, conversational tone that disarmed most callers. That sounds like fascism to me, she'd say, or: Have you lost your mind? She bypassed stickiness, remained sweet but firm, and viewers seldom raised a sound in protest. She had worked in the tough Brownsville section of Brooklyn as an investigator and case worker for the welfare department. She dealt with extremes of poverty and of human behavior the rest of us had chosen to overlook. She had fewer illusions than the rest of us, too. She won respect because the moment you met her you knew she didn't care whether or not she got your respect as long as you gave her space. The evening I first met her, she wore an irresistible miniskirt, and David wondered later, could he and she ... Naw, I said, I didn't see it happening. Fortunately for me, neither did she.

Sam loved to see her on the air and would invariably be the first caller whenever she was host. He'd inquire about her health and the family and everyone else in Videofreex. It didn't matter if he'd left the farm five minutes earlier to rush home to his house to watch the show; he'd call and ask. "Ah, Mrs. Teasdale. What? No that's not your name?" He had trouble with the concept that she'd retained her own last name. "Wai' a minute. Awright. You're mine friend, Kaaarrell."

David would host the show when he had something to say, although he, like Annie and Davidson, needed coaxing. Odd, with his personality. One summer he decided to become an expert in edible plants, and he and a friend planned a whole show on the local flora they'd found and cooked. All during that show, which we shot on the front porch because it was cooler there and we could invite more people to watch the production, a big pot of water filled with plants and leaves simmered on the stove. By the time the show was over, the concoction had been reduced to a foul smelling, purple gruel. The odor made me gag, but David insisted it was the most healthful food ever cooked at the farm. Most of us refused to eat it. Those who did try it, including David, became violently ill. Or maybe I'm suffering an olfactory flashback. Regardless, we did no more shows on the subject of edible plants.

By and large, the Lanesville audience did not seem to appeal to David. Perhaps it was too small a stage. It had been his connections and politicking that had won for us the first big grant from the State Council on the Arts, but now he was restless. He couldn't adjust to Lanesville TV as the basis around which to build an arts program. He felt circumscribed by the group and the fears that Skip and Carol expressed most forcefully for the rest of us that he might suddenly commit us and all our resources to a project on which we had not been consulted.

David had few personal possessions other than Oberon. A good thing, too, because that bird would have trashed anything of value. During our second winter in Lanesville, Oberon died, possibly from the cold drafts that prowled the house. Like David, the parrot behaved in loud and unruly ways. For some time after the death of his bird, David was unusually subdued.

∞∞∞∞∞∞

Unintentionally, Lanesville TV had become a sort of litmus test for the group. Those of us who saw it as a central function of the Videofreex and Media Bus comprised one side. David and Davidson, who viewed the station as a novelty aspect of a larger and more diverse arts center, were on the other. Plenty of projects still united us. In the fall of 1972, for instance, we spent nearly a week at the Fenimore House in Cooperstown training museum curators from around the state how to use video in their museums and their communities, and ending with a hit-and-run Lanesville TV-type program on the Cooperstown cable TV system. Earlier that year, we had packed up all our equipment and taken it to Shea Stadium to produce the video projections for the Newport Jazz Festival's concerts, wandering around the stage with entertainers like Roberta Flack and Stevie Wonder. When we weren't working cameras or switcher, Carol and I took turns minding Sarah in the Mets' bullpen dugout, shooing away the vicious mosquitoes.

And other, smaller projects also drew us together as a group. One of our favorite games in the city had been to split the screen, with the top half of one person's face matched to the bottom half of the face of a different person, or the left half of one faced matched to the right of another. The effect was fun and simple to achieve. One day I noticed that as Chuck was repairing a camera he inadvertently switched two wires and the camera produced a mirror image of reality–everything was reversed. That evening, he and I set up a system in the control room with a reversed camera next to one displaying a normal image. By pointing them at exactly the same spot and positioning a person on that spot, a split screen gave a person two left or two right sides mated in the middle. The effect was at once bizarre and hilarious. As more Videofreex came out of their rooms to see what had caused all the whooping, we began to play with the possibilities; the resulting images became progressively weirder. The taping went on well into the night.

These playful sequences were edited together by Nancy and Bart into a short tape she called "Oriental Magic Show with a Man in a Box and a Barbarian", Davidson's title for what had started out as quite a different type of tape. The final edit had a sound track that mixed our voices as we played with an upbeat piece of traditional Chinese music. It might have stopped there except that David had become intrigued with the idea of removing the control of video images from the person behind the camera and extending that control to the person appearing in front of the camera lens. This appealed to his ego, because he preferred having the maximum possible control over his own image, and, as a generous person, he wanted to extend that same measure of control to others. So he arranged a gig for us to play "Video Games" like the ones used in "Oriental Magic Show" at a small arts center in Rome, New York.

To promote the event, we went to Synapse, the student video center at Syracuse University, where we produced a full color tape of the freaky images we'd begun to experiment with at the Farm. The arts center connection gave us entree to a small commercial TV station in nearby Utica, where, after protracted negotiations with a skeptical management, we secured a half hour of unsponsored air time to play the tape from Syracuse and to use the broadcast to demonstrate the types of games we planned to offer at the arts center.

At first the station's crew reacted warily to the sight of a bunch of longhairs running around their station, asking them to do things they'd been trained to regard as mistakes–two faces superimposed, etc. The atmosphere improved when a dignified looking man with graying hair admitted he had just enough time each day to wipe off his Bozo the Clown make-up, change his clothes and run from the kids show set to the news desk to do the noon news. It sounded like Lanesville TV, and we suddenly found we had plenty of stories to share.

"Wait'll the boss goes home," a cameraman whispered to me. That evening, he went to his locker and pulled out a reel of videotape hidden there. "This is our stuff," he said proudly as he showed us goofy things he and his buddies had experimented with after hours at the station. No one other than the guys at the station had ever seen it before, but they wanted us to know that even though they were union men working at a conservative station in a small town, they were with us in spirit.

The arts center gig went off smoothly, but judging from the wary response from a few adults, some elders of mid-state New York were not yet ready for Videofreex, let alone freaky images of themselves on a TV. They had come to regard television as the one place certain never to offer anything out of the ordinary. Young people and kids, however, flipped out and often had to be dragged away. This mixed reaction did nothing to dampen David's enthusiasm for in-front-of-the-camera-control systems, and he went on to devote years to working new, always more sophisticated and cleverly disorienting projects in the same vein.

Video Games was the last exclusively Videofreex project other than Lanesville TV that involved all of us in one way or another. It was ours from beginning to end and we shared a vision that translated easily into games anyone could join. We could point to the message embedded in the games, that the medium did not have to be controlled by remote powers intent only on making a profit, a message made more palatable by the fun people had as they played.

13
Counting Each Dime

Once in a while we'd meet a Lanesville resident who'd say: No, I never watch your show. Talk a little longer, though, and something would slip out, like the two elderly women who went on about how cute Sarah looked as in infant in her bassinet. They could only have seen that image on the station. We called these folks our closet viewers. We had no way to measure the size of our audience, to test our assumption we were reaching a sizeable number of closet viewers, people who fussed with their dials on Saturday nights, tuning us in, or tuning us out. Some people refused to watch on principle as a boycott of hippie scum, or worse, boring shows. But of the few hundred homes within the reach of our signal, we guessed those who never, ever watched lived too deep in the hollows or too far behind the folds of Hunter to catch a bounce or a bounce of a bounce of Channel 3's transmissions. Within a few months of going on the air, the comments we picked up in casual conversation let us know our neighbors were aware of us and, more or less, what we were up to.

The tape of the trout stocking and others like it, many of them unexceptional compared to the elaborately edited and planned productions that characterized the best of the station in its heyday, established Lanesville TV as a serious presence in the community. One early show featured two tame timber wolves brought to the Farm by their handler. He explained the contribution wolves make to the ecosystem and how they were being hunted to extinction for no good reason. This was new information at a time when wildlife shows on TV shied away from overtly political stands on environmental issues. We broadcast a short film of his about men in Alaska who shot wolves from helicopters. The stark images of slaughter incensed some local nimrods. We got calls condemning the practice, just the type of awareness we hoped to generate.

The wolves were huge, much larger than most dogs, and extremely docile. When word got out they were at the Farm, people from all over the valley, many of them kids, showed up to see and pet them. The wolves calmly and with great dignity endured all the squeals and hesitant patting from these strangers. Our first-hand familiarity with them made it all the sadder when we reported a few weeks later that somebody had slipped poisoned meat into their cage in New York City. Both those beautiful, intelligent creatures were dead. It was a troubling and unexpected way to drive home the irrationality of our fear of predators in the wild.

Word of mouth advertising drew people to the wolves, making the audience for that evening's show as large as any we'd ever had. These initial programs opened up Maple Tree Farm to the community and simultaneously served as our entree into the lives of the people who lived in Lanesville. Like most members of the media we assumed to ourselves the right to report on the doings of our neighbors, and our neighbors, to varying degrees, accommodated us.

There was a certain language barrier to be overcome at first. The people of Lanesville might mention something about what went on "up to the Videos," as the Benjamins used to say. But they never referred to us using the name by which all our colleagues in the video world knew us: the Freex. I attribute this to a combination of their unfamiliarity with hippie jargon and their sense of politeness. We had rejected the trappings of our middle class origins and reveled in our outwardly unconventional appearance. We hoped our neighbors saw in us a reflection of the same independent spirit that made them outsiders, too. This unreasonable expectation raised a tricky question: If we were freaks, did that also mean they were some sort of freaks? Unlike us, most of our neighbors had not embraced the role of outsider by choice. They endured labels like hillbilly and woodchuck, terms of scorn applied to native Lanesvillians by residents of metropolitan Tannersville and Phoenicia. But freak was too deprecatory a description for the people of Lanesville to accept. They lacked access to the cultural power and self assurance that immunized us from the hurtfulness of a word like freak; we had appropriated the term for our own ends. Pressing our neighbors to call us the Freex would have alienated us from the audience we wanted to reach. So while we never specifically proscribed the practice of referring to ourselves by that name, we gradually dropped it in favor of identifying ourselves locally as people from Lanesville TV.

As representatives of this new entity, Lanesville's own TV station, we had something to offer. In fact, we gave it away for free. Not so strange, when you consider that the absence of cable TV or satellite reception meant no one in Lanesville had any experience with paying for television, once their antennas were installed. That may explain why none of our neighbors ever questioned why we didn't charge for what we broadcast or why we didn't seek some other form of local support.

If we had tried to sell advertising, which we never did, the two bars, Doyle's and Babe's up the road in Edgewood, the general store and maybe, stretching it a bit, the sawmill, were the only retail businesses in town, and could hardly have supported a station, even one as low budget as ours. That's assuming they'd have wanted to advertise in the first place, which assumes a lot.

Perhaps Doyle's and Babe's competed for customers. More likely, when a patron was ejected from one, he or she hung out at the other watering hole until the suspension period ended. The two bars looked much the same, long, dark interiors, with open floors large enough for square dances. Doyle's had the added attraction of booths along one wall. Jerry tried hiring local rock bands one season. It didn't draw many people and some of the regulars complained about the loud music and all the longhairs who only bought beer, if that. The music ended as suddenly as it began.

Babe Santora owned Babe's. Her raspy voice was filtered through an accent that sounded as if she hailed from the same neighborhood in Brooklyn as Jerry Doyle. She wore a beehive hairdo and heavy make-up, and she squinted at the workings of the world through a haze of cigarette smoke. She always treated us cordially but with a hint of suspicion. Carol and I went to one square dance at Babe's Bar on an evening when the Benjamin and Neal clans were well represented. We had a good time laughing and trying not to bump into anybody while we do-si-doed. Gene and Gert made a special point of befriending us, and Gene danced with Carol, swinging her around the floor, just barely maintaining his balance. The square dance was a family affair and the drinking there, as it was for Doyle's bands, was minimal by Lanesville standards. The dances, too, ended abruptly and were never revived.

When we first moved in, there was no other general store in the Stony Clove Valley but the one in Lanesville. The store in Chichester had closed after the woodworking factory shut down, and though a number of people had tried to make a go of it over the years, each attempt ended in failure. Chichester was too close to Phoenicia, which had several small markets, a hardware store, gas station and four bars. No one in Phoenicia nor in Tannersville or Hunter depended on business from Lanesville, so the neighboring villages, where Lanesville TV was virtually unknown and could not be received, would not have been fertile ground for advertising, either.

We could have adopted the public television model and begun soliciting memberships, but only a mere handful of our most diehard viewers would have been willing to contribute, and whatever they could have given wouldn't have made much difference in our finances–sort of like Davidson standing by his upturned hat at the doorway of the Prince Street loft.

The only person who tried to use the station for commercial ends was "Rabbi Kelly," who doubtless would have balked had we asked him to pay for it. One Valentine's Day he called the show to ask Nancy about an elderly woman singer she and Bart had taped at the condominium where Bart's parents lived in Florida. The woman's reedy renditions of old chestnuts had been the main tape on the show that night, and Joe, as someone who'd once been in the singing business, was checking to see if maybe he knew her. Without so much as catching his breath, he segues into a plea for new tenants: "I got six rooms for $170 plus utilities ... ."

Nancy, a little out of it, misses that last part. "No facilities?" she asks. At Joe's place, it seemed altogether possible the plumbing didn't work.

"Plus U-Tilities!" shouts Joe.

Whenever we taped Joe, he'd plug his rooms or try to sell his house. He barked more than spoke, and I never saw him talk to Bertha that he didn't command her. "Shoena!" he'd snap, expecting her to drop what she was doing and come to him. Shoena means dear or pretty in Yiddish, but it didn't sound endearing the way he said it. He'd look right at the camera and no matter what you asked him, he managed to slip in, "I got a wonderful house here and sixty acres I could sell if only I had a buyer. You should tell somebody to buy my house ... ."

A Lanesville man who swore he'd witnessed the scene told me Joe had once agreed to sell a small parcel of property he owned on Diamond Notch Road. Joe wouldn't compromise on his asking price, and told the buyer in no uncertain terms to pay him every dime up front before he'd hand over the deed. The buyer became so incensed he decided to take Joe at his word. He arrived with the payment in the back of his pick-up, all in dimes. He demanded his deed, but Joe told him he'd have to wait. The buyer, flabbergasted, asked: What now?

First, said Joe, I got to count the dimes.

No matter how Joe Keley wanted to use the station, we didn't really expect to charge our neighbors for our programming. We considered ourselves a public service paid for by their tax dollars, a service they deserved. By not asking for money from anyone but the state and federal governments and other remote funding agencies, we had left ourselves free to broadcast whatever and however we wanted. If that pleased our audience, fine. Anybody who didn't like it or couldn't receive the channel that night could look for something else to watch. The only counterbalance to this freedom was our own implicit agreement among ourselves soon after Lanesville TV went on the air that it was important for us to reach out to the community, to tape anything for anybody in Lanesville who asked.

A key element of this video covenant was our implicit agreement that we wouldn't go out of our way to offend the viewers. It happened occasionally, as when Peggy Harley from next door, who sounded as if she'd been drinking, mistook Skip's dual-sided armpit in the *Oriental Magic Show* tape for full frontal nudity. "Oooohhh, Peggy," Nancy cooed into the phone, "that's your Rorschach test for today." If other viewers had the same misapprehensions, they didn't call to complain. We never did present full or even partial nudity on Channel 3 that I recall. But here was a real dilemma: How could we claim to be a self-respecting pirate TV station, the first anywhere that we were aware of, while acknowledging one of our most ambitious early pirate broadcasts was *Lanesville Country Kitchen*? Our solution was to abandon any local talk of pirate TV, just as we'd muted our use of the word Freex. Some Lanesville people logged, worked at housekeeping or drove a truck. We ran a TV station.

∞∞∞∞∞∞∞

It's not as if the community gathered us to its bosom–it hardly had one to gather us to. Davidson was the only one of us who hunted and fished, and he chose not to join the powerful Rod & Gun Club. Nor had any of us visited the Methodist Church at the time the station signed on the air. The local bars offered no obvious enticements for socializing. So prior to Lanesville TV, our only entree into the lives of our neighbors came through Sam.

Chuck grumbled about Sam showing up every day. He said Sam disrupted his train of thought; and why did we want a nosy old man hanging around, anyway. Chuck needed to concentrate on keeping our cranky gear running and on his own projects aimed at coaxing our low-cost video equipment to do things its makers never intended it to do. Sam was a distraction for him. But I have a short attention span and am readily drawn to the loudest noise around, especially if it's conversation. Sam and I found each other naturally.

He told me he was born before the Russian Revolution in the Ukraine in a village he pronounced "Tsipanuk," which I have never found on a map. He said it meant "Chicken." He considered himself a Russian Jew, not a Ukrainian. He told me his village was filled with "Jewish guys,"–Hassidim–for whom he clearly had little tolerance. As a very young man, he traveled to Kiev to see his hero, Lenin. Eagerly I asked him what Lenin had said and what it had been like to see the leader of the revolution. "I don' know," said Sam, frowning. "I get there a day late and Lenin is already gone."

Civil war gripped Russia following the revolution, and Sam found himself conscripted into the counter-revolutionary White Army. He would have preferred to fight for the other side, but as a peasant in the midst of an epic struggle that stretched across a continent, he had no choice. As it turned out, he ended up fighting Poles, not Bolsheviks, and was wounded in battle. That ended his military service. "The Polish peoples I don't trust," he told me.

Miriam was born in Poland, but she didn't factor into Sam's blanket mistrust because he met her after he came to America. His attitudes about people and their origins had everything to do with his belief in this country. Sam lived the ideal that once you arrived here and became a citizen you were American regardless of where you came from. He hadn't abandoned old animosities or stereotypes but he didn't apply them reflexively. Everybody got his chance in America.

Sam landed in Baltimore in 1920 or '21, worked there as a laborer and then made his way to Brooklyn. Other members of his family came over, too, a brother, Alex, and Bertha. During the Great Depression, he worked as a carpenter. He always made a point of noting his lack of skills, saying, "I'm a rough carpenter, a laborer, a woiking man." He considered himself a proud member of the proletariat, and we talked politics hours on end.

"Come the revolution ..." I'd say to provoke a conversation.

"Oh Meester Parry, not going to be revolution in this country. We don' need it."

When I knew him better I asked if he ever joined the Communist Party.

"No," he said slowly. "Never I was a member of the Party. I'm a Democrat and a socialist."

I wasn't satisfied with that answer. He knew a great deal about the left wing of the labor movement. He was an unshakable union man. He owed his pension to the union, as well as the free health care for himself and Miriam, who wasn't well. Thoughtlessly, I pressed him. I had a romantic notion he might have a story to tell about a real underground movement here in the U.S. He said he knew plenty of party members. But even if he had been a communist–and I have no reason now to doubt his denial–how could he have told me? He had so much to protect in a nation that still persecuted communists. We were at war with a communist country. He could even have been shielding me. He'd seen what happens when the government presses people to inform on friends and neighbors.

Even before the death of Leo, the most obvious focus of Sam's affection was his German shepherd, Blackie. He called him "Bleckieboy," or simply "Bleckie," and doted on the animal in a way that might have been charming had Blackie been anything less than vicious. Sam claimed Blackie had been a sweet-tempered puppy until neighborhood kids had driven him mad by firing guns next to his ear. Whatever the cause, he seldom let Blackie off a chain. When Sam walked him down the road in front of the Farm each morning and evening, it looked more like Blackie was walking Sam. The dog tugged so hard at the leash he rasped with each breath, and Sam stumbled along behind, barely able to maintain his balance. Nobody came near Sam when he walked Blackie. Nobody but Mushroom.

A fast-talking city vet persuaded Nancy and Carol to adopt Mushroom just before we left for the Farm. He didn't tell them the dog had probably suffered from distemper as a puppy and had lost control of his bowels. So Mushroom, a sweet mutt alleged to be mostly Bouvier, lived outside, where the highlight of his day was standing on the bluff overlooking the highway, barking madly at Blackie and Sam. Blackie would snarl and rear, nearly escaping Sam's grasp. Sam would yell at Mushroom and shake his walking stick. You could set your watch by the commotion, and we learned to stay out of sight during these encounters.

Sam could never pass up an opportunity for conversation, and one day when Mushroom had gone off somewhere, he headed up the driveway to see me. Blackie's head was down and his ears back. "Say, Meester Parry ..." said Sam. Just then Mushroom came shooting from the back yard. He leaped past me onto Blackie. Sam tried to yank Blackie away but the two dogs were biting and snarling in what looked like a fight to the death, one that Blackie definitely was winning.

I shouted at Sam to leave, but this turned out to be the wrong approach. Sam stopped hollering at the dogs and sputtered at me. "See here! You got to put your dog auf de chain!"

I couldn't reach Mushroom's collar without placing my hands closer than I dared to Blackie's powerful jaws. So I grabbed Mushroom's hindquarters and I pulled him beyond Blackie's reach, at which point he pooped on my feet.

By this time other Freex had run outside. The shouting and barking escalated until Sam retreated, tugging Blackie, who stood on two legs, frothing at the mouth.

Sam walked Blackie up the road away from Maple Tree Farm after that, and our friendship cooled for a day or two. But not even Blackie could dampen our bond with our landlords, who seldom asked for more rent or complained if a payment was late.

From our vantage point, it appeared the Benjamin and Neal families comprised the core of the valley's population. Actually, a family named Lane were the earliest white settlers of record, arriving in 1816, two years before the formation of the town of Hunter. And there were other old clans remaining, like the Crosbys, whose forebear had built the Farm. A company called Neal & Norcutt set up a chair factory in the mid-1800s at a place called Nealsville, a settlement up the road from us that does not appear on later maps. By the 1880s, the two-story factory was owned by J.V. Neal and Sons, employed 15 men and turned out nearly 15,000 chairs a year. The sturdy, ornately milled chairs around the kitchen table at Maple Tree Farm may have come from that factory or one nearby.

A few years before we got to Lanesville, the last of the Stony Clove furniture factories closed. The Schwartzwalder plant in Chichester, a rambling conglomeration of grim, utilitarian buildings in the best tradition of nineteenth-century mills, still stood abandoned and rotting. Private citizens now owned houses in what had been the factory village. Turn-of-the-century photos preserve a practice session of the factory's brass band. There was an indoor basketball court. The boss's private pond and lodge lay beyond a stone arch in a hollow above the factory. The good manufacturing jobs left the area altogether with the end of the Chichester factory. Chuck found scraps of maple at the factory and put a beautiful new floor in the music room.

The physical evidence of vanished commerce was evident all around us. We could see it in the web of logging and quarry roads that threaded the mountainsides, the outlines of ponds that collected streamwater to run the sawmills, and the overgrown railbed.

The first trains tootled up the valley in 1882 on narrow gauge tracks. Passengers for the boarding houses of Lanesville or the great hotels overlooking the Wall of Manitou changed at Phoenicia to the mountain goat Stony Clove & Catskill Line. The train puffed up the valley past fields and farmhouses. Fields! Hard for us to imagine in the midst of the forever-wild Catskill Forest Preserve that no forest preserve existed at all until 1885. In less than a century, the Catskill preserve had grown to almost ten times its original 34,000 acres. Hardwoods had reforested slopes stripped bare for the hemlocks, whose bark fed the once thriving tanning industry. Evidence of farms in the Stony Clove had disappeared but for houses. Swatches of yellow paint slapped on tree trucks marked the boundaries of the preserve, and woe to any logger who cut his way into state land.

Standing in the front yard of the Farm in winter, we could make out the railbed through the bare trees. It looked like another old logging road. Plenty of people had found work on the railroad or a livelihood that depended on the line. Old man Crosby would never have built Echo Cottage without the trains to bring him guests.

At the notch, over 2,000 feet above sea level, the rails intersected the wagon road, and the line employed a resident attendant to flag the crossing. The cabin where he and his family lived lay nearby, and late one winter, while his wife sat in her rocking chair in front of the coal stove nursing their baby, a passing train shook loose a boulder from high up on Hunter Mountain. The great rock roared down into the notch, burst through the cabin wall and exited the other side, removing the stove but leaving mother and baby unharmed.

I first heard that story from Willie Benjamin, the man we considered the patriarch of the valley. Willie had worked on the railroad as a young man. His daughter, Gert, married David Neal but hung out with Gene Grant. Willie's son, Bobby, married a woman named Harriet, the mother of Sheriff John. She had grown up in Connecticut, an outsider, like us. Willie lived in a small, two-story clapboard house diagonally across the road from the general store. Bobby, Harriet and their four sons and a daughter lived just up the road in a trailer they'd expanded in several directions to accommodate their brood. David and Gert Neal lived in a small, ramshackle house just up the road from us, two houses north of Sam and Miriam. Gene lived in a bread truck in the meadow above them. Willie's brother Elmer and his family lived further up the valley in a compound of small houses, and his brother Frankie lived in a tidy roadside home in Edgewood.

Willie had trouble walking because of arthritis. He used a cane, sometimes two of them and sometimes crutches to move his heavy body between his tractor, his dump truck and his Cadillac. He didn't have many teeth and the sharp features of his wide face cascaded toward his mouth. He smiled quickly and cackled when he laughed.

Elmer, a gaunt man who shared Willie's etched face, fathered seventeen children. Shortly after Lanesville TV went on the air, Carol dropped by to interview him. She didn't approve of such large families, but her manner remained genial. Seventeen children, she remarked. Can you name them? Nancy, behind the camera, gives out an audible gasp. Elmer looks befuddled. "Well," he says slowly. He pauses for a long drag on his cigarette, "there's Elmer Jr. and Bobby and ..." He pauses again and shakes his head.

"Charlotte!" comes a loud, off-camera whisper.

"Yeah, Charlotte ... ." And he rattles off a few more. With additional prompting from children and grandchildren he gets most of them, but the tape does not disguise his struggle to remember. He and his family presented a good-natured face while Carol and Nancy shot their tape, but Elmer's branch of the family, unlike Willie's and Frankie's, seldom had much to say to us from that point on.

∞∞∞∞∞∞∞

After our noisy behavior drove the elderly people from the house next door to the Farm, the place remained vacant for a while. It, too, was the product of David Crosby's labors. He built it to accommodate the overflow of summer guests. Later, when the summer business dropped off, his successors had sold it as a separate home. It needed a lot of work. The next summer it was sold to Neil and Peggy Harley. They had a daughter and three sons. Neil Harley's broad, red face was a prizefighter's nose governing other, smaller features. The boys had more hair and far less ruddy complexions. All the men in the family worked as sandhogs in the city, building tunnels. As soon as they bought the house, the Harley men went to work with shovels and an immense quantity of

Budweiser—the high-priced stuff by our standards—and proceeded to excavate their basement. They spent weekends underground, while a tinny speaker in the window blasted rock music they could hear as they wheeled out small mountains of dirt. They returned to the city each Sunday night.

I never saw the basement up close, but I had an uncomfortable feeling they'd extended it under the yard between us and that any weekend they might spill out into our cellar. The senior Harleys were friendly, and Peggy was a big fan of Lanesville TV. But their children's insistence on sharing their music with the neighborhood coupled with the whole family's bouts of drunken shouting grated on us. Before long we began to resent the Harleys as loud weekenders and outsiders. Then the Lindseys arrived.

Maple Tree Farm shared the small bluff above Route 214 with two other houses. The Harleys' house lay about thirty yards from the north side of the Farm. Close to the other side of the Harleys was a modest, two-story frame dwelling. It was unoccupied when we arrived and remained empty until our third winter at Maple Tree Farm, when a family moved in. We didn't meet anyone from the new family right away, but now that we had a better sense of Lanesville, mysterious neighbors no longer seemed odd. Carol made first contact when she took Sarah for a sleigh ride. She met a tall, muscular man, whose tracks through the snow indicated he'd just walked down the driveway that led from the house to Neal Road. Carol asked whether she could use his driveway for sledding. He agreed and they got to talking. When he asked about the Harleys she mentioned that our mutual neighbors drank a lot. "I shoot heroin in my eye," replied Ray Lindsey.

Ray owned a chopped Harley, which he kept in his living room. He had adorned his body with enough tattoos to qualify for a sideshow. He claimed to have friends among the Hell's Angels chapter in Manhattan, though he let us know he didn't want official membership of the club. Not a joiner, and maybe a little too rough around the edges to qualify.

Years before, David and I shot a tape at the Hell's Angels' basement clubhouse in Manhattan. They had a slick, young leader named Sandy, who saw us as a way to garner good publicity for his version of the Angels: just group of guys who want to ride motorcycles and have a good time. He invited us to a tape a party the Angels were giving at a private, midtown Manhattan club, a surprisingly tame affair meant to show the mild side of the notorious gang. The only memorable sequence is an Angel named Filthy Phil leaning drunkenly into the camera to impart the pearl of wisdom, "Just because you don't tell the truth don't mean you're lyin'." Our last encounter with the Angels came a few weeks later, the night we were supposed to play tapes for another Angels' celebration, this one at the Electric Circus in Greenwich Village. It was a major public relations coup, a public event at which the Angels would demonstrate that, hey, we're not violent or criminal like the media says. Come party with us. Just before the show, Skip and Alan Sholem stopped by David's loft on Rivington Street. While they were inside, junkies made off with all the video equipment in the car. Skip and Alan hoped the Angels would put out the word and get us our equipment back. But Sandy was furious. He swore at Skip and Alan, and made it clear they weren't welcome without the tapes and the machinery to play them for his guests. We didn't tell Ray this story, only that we knew Sandy.

Ray wasn't real talkative, but he managed to communicate fairly well. One summer day he parked his bike in our driveway and began polishing it in anticipation of leaving for a biker rally somewhere in the unspecified "north." The Grateful Dead might show up there, he said. Man, did he dig their music.

I said I didn't care much for The Dead.

A long silence followed. Slowly, he stopped polishing his ape hanger handle bars and looked up at me, glowering. Our conversation had ended.

That incident aside, he and his biker friends considered Videofreex to be enough of an outlaw organization to meet the threshold criteria for acceptance. We had long hair, liked rock music except when the Harleys played it to accompany their subterranean activities, and shared a mistrust of authority. Knowing Sandy and some of the other Angels in the city confirmed our credentials.

Little of what we knew about Ray Lindsey came directly from him, his wife, Kathy, or Kathy's young kids from a previous marriage. Our source of information was Ray's brother-in-law, Richie. Richie and his wife, Maureen, rented the cabin behind Maple Tree Farm the summer after the Lindseys moved to Lanesville. Richie was Ray's Boswell. He could barely constrain himself when it came to telling and retelling the legend of Ray Lindsey; how, for instance, the Angels in the city had loosed their German shepherds on Ray during one visit, just as a joke, and how Ray had silenced the dogs, knocking each of them out cold with one punch while not suffering so much as a scratch himself. "You shoulda fuckin' seen it, man," said Richie. "The way he fuckin' knocked out the fuckin' dogs!"

The summer Richie arrived was also the summer state police arrested Ray at gunpoint just up the road. He'd been recognized from a wanted poster at the post office. They extradited him to New Jersey. "He stomped a coupla troopers down there before they was gonna stomp him," said Richie. "He broke one of their fuckin' legs and smashed the other one's nose off, but they don't wanna fuckin' admit he coulda done that much damage to 'em, them being troopers an' all. So he'll be out soon."

Richie had that last part right. It didn't turn out to be good news for him, though. Richie'd had a fist fight with Maureen on the lawn of Maple Tree Farm while Ray was away doing time. Richie was no Ray when it came to fighting, and I scored the bout in Maureen's favor. Despite giving better than she got, she reported the fracas to her sister, Kathy Lindsey. When Ray returned, he explained to Richie the consequences of ever touching his sister-in-law again. Shortly thereafter, Richie and Maureen moved out.

The Harleys acted as a buffer between us and the Lindseys. Their two houses were set close together and a lot of shouting went back and forth. Peggy Harley swore one time that Ray jumped through a closed second floor window in an effort to catch up with one of her sons he accused of insulting him.

The new chief of police in Hunter, Jerry Gerrard, showed up several times. Like Ray, he was a Vietnam veteran, said to have reenlisted twice for tours of duty as a doorway gunner in an army helicopter. Gerrard's law enforcement methods allegedly included telling Ray he'd meet him in the woods and they'd settle the matter between them if Ray didn't behave himself. The tactic may have worked for a while. But one particularly wild Labor Day, we heard gunshots from the Lindsey house and the police were summoned again. After that, the family moved out and we didn't hear about them for years.

14
The
Lanesville
Players

I suppose it's fair to say The Lanesville Players started with David and Bart's *Three Billy Goats Gruff*, although we didn't coin the name Lanesville Players until well after that tape had passed into recycle heaven–gone forever. By the same standard of attribution, *Three Billy Goats Gruff* actually started with Pedro Lujan, a Soho artist and a good friend of Davidson, David and Bart. Pedro had decided to build a bridge over the Stony Clove Creek. Far from a public works project, Pedro envisioned the bridge as art.

We numbered Pedro among our most welcome guests. He had a bubbly enthusiasm and a quick laugh, and he lacked the pretensions that accompanied so many of the people in the Manhattan video art scene, preoccupied with comparing the size of their grants with what others got. Pedro had no direct involvement in video; he built his bridge as part of a collaboration with David, who had decided to focus his attention for a while on documenting crafts.

The State Council on the Arts was funding grant applications in 1973 from individuals who proposed videotaping historic crafts. The Council wanted a record of disappearing skills and the people who practiced them. The money for the program amounted to only a tiny fraction of the more than one million dollars earmarked annually in the early 1970s for the Council's Film, TV/Media and Literature Department (video never rated its own department); but video projects of that era were often scaled to the money available. The scarcity of funding sources meant that if you didn't adapt what you wanted to do to the priorities of the Council, you weren't likely to get any money to do it at all. David, after years in the grant-getting game, knew what the Council wanted to hear. He followed closely the list of projects that received positive recommendations from the panels of artists and arts administrators who reviewed grant applications for the Council. He understood the panels made their choices based as much on horse trading as merit: I'll vote for your friend's project if you'll vote for mine ... .

The politics of this game went right over my head for a while, as I naively clung to the notion the Council awarded grants strictly on the quality of the project–whatever that meant. The Council's record of support of the video movement at a time when no other agency would consider the medium stands as a remarkable achievement. It helped launch the careers of young artists and provided access to the means of TV production to people around the state who might otherwise have been excluded from the medium altogether. But the Council had its limits, and plenty of talented people with good ideas got no money. David grasped from the outset you had to work the system, even for the strongest proposal.

He also knew it didn't matter if you deviated from what you said you'd do in your application as long as you came up with a product at the end that could be added to the list the Council used as a justification for additional funding the following year. So maybe Pedro wasn't a grizzled artisan from a Catskills hollow, crafting artifacts just like granddaddy. David discovered only one person who actually fit that description, a man named Harvey, who still carved grain shovels from blocks of wood. With Harvey as the showcase tape and in the absence of competing candidates, David didn't have to stretch the terms of the grant too far to include Pedro's project. Pedro, after all, planned to build his bridge out of rope.

Grants for special projects such as documenting crafts seldom amounted to more than a couple of thousand dollars. But one of the qualifications for receiving state funds required the recipient to show that the state was not the only contributor. We never had a problem with that. We would simply list an arbitrary amount in the application as a contribution from Media Bus. It didn't seem to matter that Media Bus received almost all of its funding from other state grants. You played the bureaucratic game by its rules, not logic.

We had no problem applying a flexible interpretation to the Media Bus contribution on our grant applications. But the issue of the group's share of the crafts tapes rekindled the wrangling that had so bitterly divided us over the Jerusalem project. Was David once again about to spend more than the grant amount without sharing any of the spoils?

David was growing weary of our suspicions. What we believed were matters of collective survival, he saw as petty and unfounded carping. He was drifting apart from the rest of us. His father's health was rapidly deteriorating and the old man and his plumbing supply business in Boston tugged at David for his attention. Both David and Davidson began spending long periods away from the Farm, though neither was ready to announce an official departure from Media Bus, regardless of how unappreciated they felt. They had invested heavily in Videofreex, not only money, but time and emotion. Shedding an identity as one of the Freex, abandoning the power of numbers and the history and the reputation so you could be just another individual artist or producer, was a daunting prospect. Practical considerations intruded, too. By then, our original equipment had become obsolete or battered beyond usefulness. Everything belonged to Media Bus and couldn't easily be sorted out and transferred back to an individual because of the nature of tax laws governing non-profit organizations. David, in particular, couldn't afford to go anywhere else and get the same types of services available for free at Maple Tree Farm. He and Davidson stuck around half in, half out.

In the summer it didn't matter. The summer assuaged the tensions between us. Maple Tree Farm seemed as large as the Catskills, plenty of room for everybody. David could work as he pleased outside. By summer we had usually received partial payment on that year's media center grant from the Council. We all had projects of our own to work on. We could buy large quantities of food and Annie would usually take time out from her work to cook dinner. The number of visitors always picked up in the summer, friends and strangers arriving from all over the world eager to see how we worked and to work with us. Pedro worked on his bridge almost unnoticed amid a flurry of summer activity.

The finished span was worthy of Whiz Bang Quick City, the week-long gathering of people engaged in designing and building alternative structures. WBQC had drawn hundreds to Woodland Valley in May 1972 to celebrate domes and yurts and tepees and all sorts of joyously inventive structures. Maybe WBQC inspired Pedro, though his bridge, built a year after WBQC, followed a traditional plan, harking back to his Latin American roots. It had rope sides made of thick, bristly hemp, with small logs lashed in place for a walkway.

On its swift course through Lanesville, the Stony Clove Creek seldom reaches a width of more than a dozen feet, except where the land flattens for a short distance in a glacial outwash near the county line, or upstream as it burbles through the remnants of mill ponds. The water never runs more than four feet deep, with many sections two feet or less. Pedro's bridge spanned a shallow part of the stream directly across Route 214 and down an embankment from the Farm, at a spot where, a few at a time, we drifted over to cool off on unusually hot July days. The bridge swung gracefully between trees on opposite banks. It blended with the surrounding forest. But however graceful it looked at a distance, when anyone tried to navigate it, the rope stretched and the whole bridge swung unsteadily side to side. Everyone who attempted a crossing became an ungainly oaf, and people on either bank would laugh and laugh and laugh until their turn came.

When the Benjamin kids found out about the bridge, nothing could keep them off it. Its wobbling made attempts to cross it all the more alluring, with the penalty for failure an unthreateningly short drop to shallow, icy water.

Like the kids, David had a particular fondness for splashing in the water that summer. In conjunction with his crafts tapes, he had embarked on a nature trip, casting himself as our household expert on Catskills flora. He documented his botanical explorations on tape by partially unscrewing the lens of the camera to obtain video close-ups otherwise available only with lenses we could never have afforded. We'd picked up lots of on-the-cheap techniques like this as our budgets dwindled in inverse proportion to our ambitions. David's rambling, unscripted and questionably researched tapes offered a free-form alternative to the stiff, omniscient-narrator nature documentaries of the era. He recorded his process of discovery, which sometimes yielded insightful sequences, and just as often became painfully tedious.

David had immersed himself in this microscopic view of the world until one sunny afternoon, when the Benjamins were goofing around on the bridge, he looked up long enough to cajole them into playing his game rather than theirs. With Bart recording the scene, David crouched under the bridge and directed Sheriff John and the others to take the parts of the billy goats. Long sequences recorded David's instructions to his billy goats and their questions about script and motivation. Most of the kids knew the story, and the hold-ups had to do with their interpretation of his direction. The process grew theatrical–familiar territory for David from his life before video. Stopping and starting the VTR produced a roughly edited tape, and the dialog, such as it was, came and went in the gurgling of the stream. The idyllic setting of woods and bridge enhanced what looked like the genuine reluctance of some of the kids to pass within reach of a troll like David hunkered down in the water–his frizzy hair had grown particularly wild as had his beard, his naked torso was matted with fuzz and his voice projected deep and growly with more bluster than menace–all of which lent the tape a magical edge.

The performance transcended the obvious contrivances of technology and dramatic craft to become a found theater in which video endorsed the realness of the moment. Watching this tape and others like it made you believe the camera, as an active observer, concealed nothing, showed you everything happening at that instant, gave you back the time as you would have experienced it, which, of course, is the essence of illusion.

We played *Billy Goats* on Lanesville TV and then pretty much forgot about it, another "probe" as we called it, a trial run with no particular expectations about the outcome. As long as we had blank tapes available and charged batteries and a portapak that worked, no one cared who shot what or when.

∞∞∞∞∞∞

By the time of *Billy Goats*, we had accumulated about four portapaks, two of which could be counted on to work reliably at any one time, with another one iffy, and the fourth on Chuck's bench after having swallowed some valuable footage at a critical time. We had all learned to expect, though not accept, that the equipment would frustrate us when we needed it most.

Sony made the most reliable portapaks and that's the only brand we bought. But even the best half-inch portable VTRs couldn't withstand the beating we gave them. Chuck bore the brunt of our frustrations, with all of us at one time or another uncharitably attributing inherent limitations of the technology to a lack of skill on his part. Lighting one Lucky Strike after another on the stem of his soldering iron, he worked in a blue haze diagnosing and repairing the quirkiest of technical problems.

Chuck had an opinion, usually stated as fact, on every subject. We frequently had to check ourselves whenever we verged on reciting data for which Chuck was the source, unless the data applied to video equipment. For all the class gulf between those of us with our bourgeois upbringings and Chuck, the product of Catholic orphanages and less inviting institutions, all of us depended on him: "Something's wrong, dammit. It's not working! Where's Chuck? Chuu-uuck!! Oh, shit. Look at that. Can you fix it, Chuck? Can you!? ...."

With Chuck to back us up, we could make tapes when we wanted rather than having to forego opportunities because of uncooperative equipment. What Chuck asked in return was the privilege to experiment on projects that pushed the limits of low-cost equipment. Those experiments sometimes produced spectacular results, like his genlock or the first Lanesville TV transmitter. Other promises never panned out or came with annoying and unresolved flaws. On balance, he probably achieved a success rate higher than any commercial facility, the very type of institution that would have rejected him for his lack of education or strangled his curiosity with rules and restrictions. At Maple Tree Farm Chuck worked mostly undisturbed except for production emergencies, hunched over his bench in a cloud, hour after hour in a room that looked as if a tornado had just passed through–half-finished circuit boards piled on top of each other, books, technical manuals and schematic drawings scattered everywhere, broken cables drooping like vines over the ruins of oscilloscopes cannibalized for their parts. His shop became the dumping ground for all the stuff that didn't work, and we always had plenty of that.

Some problems confounded even Chuck's extraordinary skills, and though he seldom confronted one so daunting he wouldn't take it on, he did encounter things he couldn't fix.

We recycled *Billy Goats* because we suffered a chronic shortage of blank videotape, our raw material. And though you can reuse videotapes, as we did in the case of *Billy Goats* and many others, we preferred to use brand new tape. Recycling (recording over) old tapes meant making painful choices about what to destroy and what to save. Recycling also meant using tapes likely to have sustained some sort of damage. If damaged tape went through the VTR, it frequently resulted in clogged video heads, leaving you at the end of a shoot with nothing but blank tape. And with documentary shooting, what you've lost you cannot recreate.

We found it harder and harder to come up with the money for tape as the State Council on the Arts gradually cut back our funding during our years in Lanesville. We had become victims of our own success. The demand for video equipment we helped generate led to cheaper, more widely available gear. Cheaper equipment promoted greater access, which spurred the growth of programs like ours all over the state. We had plenty of company proselytizing the miracle of video: Besides Portable Channel in Rochester, Gerald O'Grady had established Media Study in Buffalo; former Syracuse University students had put together Synapse; and in Binghamton, Ralph Hocking, a soft spoken artist and ex-cop who looked like a cross between Santa Claus and a professional wrestler, had started exploring the most unconventional methods of generating electronic images at the Experimental Television Center. Ken Marsh and Elaine Milosh, late of People's Video Theater, set up Woodstock Community Video and were producing shows on the town cable system. Phillip Mallory Jones had created the Ithaca Video Project. The Council's money in the city had ignited an explosion of organizations like Downtown Community TV Center, the Media Equipment Resource Center and The Kitchen video theater, all in addition to an ever growing number of individual artists all vying for funds with the original groups like Global Village and us. The Council pointed to the proliferation of video artists and video outlets as evidence of the beneficial effect of public expenditures on the artform. But all of a sudden we found ourselves competing for funds against groups we helped create. In the aggregate the Council and the movement had spawned too many video mouths for one bureaucracy to feed.

Media Bus no longer offered unique services, except for our officially unrecognizable pirate station. That didn't stop our projects from becoming more elaborate, and more ambitious projects usually required more videotape. Sometimes we experimented with cheap, unknown brands of tape, but the results always produced disappointments. It often came down to a choice between good tape or our $25-a-week salaries.

Part of the problem lay with our decrepit editing VTR, which we had bought back at Prince Street with funds Davidson managed to persuade his wealthy father to "lend" us. This machine, made by a company called IVC, required one-inch-wide tape interchangeable only with other IVC machines. Like an old car, it overheated all the time, and then drifted out of alignment so that certain tapes would play back only on our machine and nowhere else, and sometimes not on our machine, either. Some of the people who came to the Farm to work with us expressed misgivings about depending on an eviscerated VTR, its guts spilling onto the console. We had no choice. We never knew when we'd need Chuck to rush in and tweak it back into service.

Because top quality tape for the IVC cost so much, we jumped at a deal to buy a case of Memorex tape. A year later, Chuck and I noticed a thin white film had formed on the surface of the tapes. It coated the IVC's tape path, gumming up the works so badly the reels stopped turning. We called it "mold," though more likely it was an inorganic substance created by an unanticipated reaction between the chemical coating on the tape and the atmosphere. Whatever the cause, the tapes became useless and the sequences on the tape irretrievable. Chuck looked at the problem, thought about it for a while, made a few calls and then shook his head. No hope. No one had a reasonable way to salvage the material stored on these tapes. I lost several pieces to that mold, gone for good. The experience soured me for years on recording work on videotape. Instead, I redirected my efforts, concentrating more on our live broadcasts and writing.

∞∞∞∞∞∞

With good tape and bad, we turned out plenty of probes, and plenty of them we intentionally erased as a waste of tape. We worked with kids frequently because they reacted openly to us, without the barriers adults erected as a conditioned response to the challenge of the camera. One Easter Nancy walked onto the Lanesville TV set as Carol interviewed several of Harriet Benjamin's children. She wore white long johns, clown make-up and a pair of rabbit ears sewed on a tight-fitting bonnet, all part of a homespun costume whipped up by Annie that afternoon. She plunked herself down in a chair next to the kids as if it was as natural as rain in the Catskills for the Easter Bunny to drop by. The kids were dumbstruck, especially three-year-old Scotty, whose mouth hung open in disbelief while Carol blithely turned to interview the talkative bunny.

The bunny became something of a tradition. Joseph Paul, who had become a regular guest after he helped get us on the air, played the role in a more cynical vein the next year, without the benefit of children on the set. I picked up the character the following season in an elaborate Lanesville Players production. The plot, if you could call it that, revolved around the discovery by the Bunny that local chickens had gone on strike–too much work to lay all those eggs for Easter. Saturday morning, wearing the bunny costume, embellished now with a top hat, a swallow-tailed coat and two giant, white athletic socks stuffed with cloth as my rabbit's feet, I wandered down to the general store to see how the people of Lanesville would react to the news. Alice may have been surprised by the rabbit. Lanesville took him in stride.

Charlie Benjamin offered his opinion on the chicken strike, as did a number of passers-by. "Rabbi Kelly" had some advice about eggs (and a plea for tenants), and no one we met even for a moment acted as if being interviewed for television by a formally attired rabbit was anything but routine. The Lanesville Players didn't have to get into character. They never got out of it.

Harriet and her kids also appeared in Nancy's tape, *Harriet*. Nancy spent hours in Harriet and Bobby's trailer taping Harriet's everyday activities. At first it seemed it would end up as a non-narrative documentary chronicling that branch of the Benjamin family, the cramped quarters, the raucous meals, the household drudgery and the demands of five kids, which Harriet managed, like most parents, with a mixture of exasperation and joy. But Nancy saw something special in Harriet, a round-faced woman with a bounce in her step and a sing-song voice that bordered on whiny except for sudden, wild hoots of laughter. Somewhere in Harriet lived a playfulness that her hard life had not managed to erase. As the two women got to know each other, and as Harriet came to trust Nancy, a different story emerged, one in which Harriet slams the door of the trailer, shouting "I've had it!" jumps in her car and drives off over the notch laughing hysterically as she experiences a newfound freedom.

Harriet was docu-drama, far less narrative than *The Cool World* or the pioneering work of film directors in Britain a few years earlier, and unlike anything that had ever appeared on TV in the U.S. No one in the American television industry could have conceived of a piece like *Harriet*. In this country, broadcasters still suffered from the shock of the radio broadcast of *War of the Worlds* in 1938. Thousands of people actually believed Martians had invaded the Earth because the broadcast sounded authentic. *War of the Worlds* exposed both the extent of public gullibility and the power broadcasting held over the American psyche. The networks would not permit anyone to attempt something like that again for decades. Certainly in 1973 no TV executive would have risked his house in the suburbs to show a piece like *Harriet* that blurred the distinctions between news and fiction for the sake of art. The networks confined their own suspension of disbelief to such items as the body count of enemy dead in Vietnam.

Info-mercials, edu-tainment and tabloid television shows still lay in TV's future, as did programming competition from cable TV. Television executives insisted on perceived absolutes of news and entertainment, as if those definitions didn't fall along a continuum of information styles. The conventional media had no idea what to make of a program like *Harriet*. That limited the size of its audience, but not the effect the tape had on those who saw it.

Harriet never did abandon her life in Lanesville, but Nancy's tape became a favorite wherever people gathered to watch tapes. It won considerable acclaim in the emerging feminist movement, with Harriet herself raised to the status of working class hero. Nancy took her to one screening at a women's conference in New York City where Harriet received an ovation for her performance, perhaps the only time in her life she had won unqualified public praise.

It couldn't have helped her standing in the Benjamin clan that she had grown so close to us, but Harriet was already an outsider because she wasn't born in Lanesville. That and her outgoing nature caused her family to ostracize her well before we arrived, which was why she'd felt free enough to work with Nancy in the first place.

∞∞∞∞∞∞

We didn't confine our probes to the Stony Clove valley. In the early spring of 1974, Bart headed over the notch to Tannersville after getting wind of plans for a local prize fight. The purpose of the fight was to raise money for the town rescue squad, but rather than bring in professional boxers, this match would pit two local guys against each other, neither of whom had a grudge or something obvious to prove. They volunteered to step into the ring together for a good cause. The names of the fighters, "Rocky" Van and Frank "The Fist" Farkle, drew Bart to Pete's Place, a tavern on Route 23A just west of Tannersville, where he planned to check out Rocky's training camp.

If Rocky had ever aspired to a career in the ring, his days of fighting trim had long since passed. His belly protruded far over his belt, and between knocking back cold beers, you could usually find him taking a deep drag on his cigarette. He blustered in an unhurried baritone made gravelly, no doubt, by his lifestyle.

On his tip-toes Frank The Fist stood eight inches shorter than Rocky. He looked nearly as round as he was tall, with a large moustache and the sharp, melodious voice of an Irish tenor. He could claim no better physical conditioning than his opponent, meaning both men had plenty of working out to do to qualify as merely unfit. Frank's arms barely extended beyond his girth.

The fight would have gone off successfully without Bart's appearance at Pete's Place, but the addition of television to the mix spun the whole event off into a new dimension. The two would-be prizefighters jumped into character with no prompting. As soon as Bart showed up to tape Rocky, the call went out to Frank–with the end of the ski season, neither man had a day job–and Frank showed up for the classic pre-fight confrontation, all of it staged but none of it planned. Bart had stumbled into a community improv theater performance, the outcome of which, as in all good improvisation, remained unknown.

Over the next couple of weeks, Bart and Nancy made tapes of the two fighters training and of people in Tannersville and Hunter speculating on the outcome of the match. Frank's mother, a slight woman with a mischievous look in her eye, said deadpan between puffs on her cigarette that her son had never been a troublemaker but that she expected him to crush his opponent. The presence of the camera egged people on, causing one normally reserved man to lean into the lens quite suddenly and stick out his tongue in a grotesque gesture Diane Arbus would have treasured. Other staid members of the community offered outlandish claims about their boy's punching stats or physical conditioning, knowing full well that both fighters were cardiovascular basket cases whose workouts consisted of lifting bottles of beer, except when the camera showed up.

All this activity increased interest in the fight exponentially, even though, with the exception of playing back the unedited tapes at Pete's bar on several occasions, no one but the people who received Lanesville TV on the other side of the notch had seen what Bart and Nancy had taped. The residents of Tannersville understood that their fundraiser had grown from a small community prizefight to a televised event. As far as most of them knew, everybody behind the scenes in the television business looked like the Videofreex.

The fight drew a standing room only crowd in Tannersville, where Videofreex had set up several cameras at ringside as the two fighters flailed away at the air and, often unexpectedly, landed punches on each other. It became a hard fought battle, with both men struggling mightily to catch their breath. And in the end, because it was not a grudge match, the judges declared it a draw, and the two fighters, who had heaped mock insults on each other right up to the bell for the first round, remained friends.

Carol and I missed the fight. Carol was pregnant again and we had made plans before we knew about the fight to visit her parents in Canada. By the time we got to Watertown, New York, not far from the border, Carol knew something was wrong. A day later, the doctor at the hospital in Perth, Ontario, confirmed she'd had a miscarriage. It happened the night of the fight, and neither of us felt up to sharing the elation of the people at the Farm over the final act of the Farkel-Van tape.

The fight raised the money for a new ambulance and the Videofreex made enough friends on the other side of the notch to ensure that, despite the conservative nature of the town, no one would hassle us there. The tape of that event remains the most spirited and accessible of all the Videofreex/Media Bus productions and a model for community programming that works.

15
Newsbuggy

Lanesville TV stuck to its regular sign-on schedule, but our connections to the community beyond the broadcasts evolved in unpredictable ways. After 1974, only our visits to the post office in its cramped corner of the former general store offered a regular opportunity to meet our neighbors face to face. Harriet invited each of us to dinner at different times, which created quite a buzz on the party lines of Lanesville. Gert brought her toddler, Dale, over to play in the wading pool. But when she discovered Carol had let Dale and Sarah, both just two years old, splash around in the pool together with nothing on, she snatched up her son, wrapped his naked body in a towel and didn't speak to us for weeks afterward for this transgression of modesty. And the summer before that we received an invitation to the Stony Clove Rod & Gun Club's annual clambake, the biggest social event of the year. It came from Scotty Stuart, who wanted us to broadcast the festivities. We hadn't joined the gun club because none of us owned a gun. So much more our shock, then, when Davidson announced he planned to go deer hunting that fall.

I went hunting once at the invitation of Timmy Hennessy, a companion of mine. We must have been about ten. Our prey was raccoon. Timmy's family considered the meat a delicacy–"It's just like chicken"–and the pelts were a major source of income for them. We started out after dark on a cold night. They loosed the dogs who howled through the darkness until they'd "treed a coon." One of Timmy's brothers trained the beam of a flashlight on the cringing animal, and Mr. Hennessy shot it with a small caliber rifle so as not to harm the fur. The carnage didn't dissuade me from hunting again so much as the tedium.

No one else in Videofreex had ever hunted, and Davidson's decision struck some Freex as politically suspect. For Davidson, though, it was a return to tradition. When he was a kid, he hunted deer with his father, and now he lived in Lanesville in the middle of the best deer hunting territory in the state. Logging roads laced the mountainsides, making access to the higher elevations easy. Clearings left by loggers offered good lines of sight. And the club had posted much of the local land, restricting hunting and fishing to members only. The signs discouraged casual hunters from stopping along the highway to try their luck.

Our objections to Davidson's hunting had nothing to do with meat–we all ate meat at the time–nor did anyone get particularly exercised over the rights of deer. To me, the thriving herd was a constant road hazard. The problem had to do with the notion of armed and all-too-often intoxicated men swarming over Sam's property. Guys wearing bandoleers of high-power shells criss-crossing their torsos would swagger into Doyle's, sidearms prominently displayed, as if they'd stepped off the cattle trail rather than driven up from Queens for a day or two. Most of us liked to walk in the woods, and we'd had a few unpleasant run-ins with hunters who refused to leave the property, even though Sam had agreed to let us post the lower acres.

Sam and Miriam liked to accommodate their neighbors. They didn't want any trouble with anybody. When they first came to Lanesville to run Maple Tree Farm as a summer boarding house, they bought a Buick with the idea that Miriam would learn to drive. She never found the time that year to take lessons, so they stored the car in a garage on the property. According to Miriam, Willie admired the Buick and offered to buy it. His price was far too low and she refused to sell. Well, said Willie, if you leave it like that, who can say if it'll still be there in the spring. Miriam talked it over with Sam, and they sold the car to Willie. They considered their loss the price of peace in Lanesville.

Our request to post the land presented them with another dilemma. In the years before we arrived, they either gave permission for people to hunt on their property or hadn't known and hadn't cared that hunters traversed the land. They valued us as tenants and wanted to accede to our wishes. But how could they tell their neighbors hunting was now forbidden on lands that had remained open for so long? They solved the problem by saying yes to everybody: Yes to us about the no-hunting signs and yes to anyone who bothered to ask for permission to hunt.

"Aaah, don't vor-ry 'bout it, Mr. Parry," Sam told me after I'd tried unsuccessfully to eject a well-armed neighbor. "They not going to bother mit you. I'm walking Bleckie every day in the woods and they not shooting me!"

I was not reassured.

Sam and Miriam's refusal to take sides left the group on morally ambiguous ground, given Davidson's determination to bag a deer. If Davidson hunted, any effort we made to enforce a ban on hunting by others would be hypocritical. The quiet disapproval that hovered around the kitchen as hunting season approached did not deter him. No use wasting breath on a confrontation. There'd be hunting on Sam's land. Davidson had already bought his rifle and a red-checked, wool suit.

In order to get his license, Davidson had to prove proficiency with his rifle. That meant a trip up to the Edgewood bar to talk with George Hatcher, who issued hunter safety certificates. Davidson didn't drive, which meant somebody had to give him a lift. He would not have hitchhiked even for such a short distance in the Stony Clove valley, where you frequently saw armed men walking along the roadside. Shortly after we moved to Maple Tree Farm, he'd been trying to hitch a ride home from someplace south of Poughkeepsie when a cop busted him, determining on the spot his vitamin pills were illicit drugs. They threw him in the Dutchess County jail for several hours until we arrived to bail him out. It was open season on hippies, and you could never tell when a cop was having a bad day or singled you out because he was bored or didn't like the length of your hair or the style of your clothes. So Davidson asked me to take him to Edgewood. I agreed, but I took some heat for abetting him in what the rest of the group considered an act of bad taste and questionable judgment.

George Hatcher had a craggy face with no trace of a smile. He looked Davidson over skeptically and scowled at me from his perch behind the far corner of the bar. At least Davidson carried a rifle, and the ice broke pretty quickly between Davidson and George, as Davidson demonstrated he could handle a firearm and knew the basics of hunting safety. At the end of the interview, George grinned ever so slightly and shook Davidson's hand. He even offered him a drink–a small breakthrough, but an important one for Videofreex. Davidson's interest in hunting made pigeonholing us as a hippie commune outside the norm all the more difficult.

Davidson met with no success in the woods that fall or any other season while we lived at the Farm. But he did shoot a deer the summer after his first outing. A friend of Ray Lindsey's heading back to the city after a weekend in Lanesville collided with a buck near the county line. The animal lay mortally wounded, in obvious distress. Someone called for Davidson, who grabbed his rifle, and, with the permission of the driver and a local cop, put the deer out of its misery. He could have left it there for the county to cart away, but someone suggested what a good idea it would be to have a barbecue. Someone else suggested the luau. So the deer was butchered and a huge party ensued the next weekend at the edge of the stream, with video people and bikers from all over converging on Lanesville. Even though the drumming and singing went on until dawn, I never saw a cop and no one complained to us. And after that, no one at the Farm had much to say about Davidson's hunting.

∞∞∞∞∞∞

The arrival of Sarah must have made us seem a little less out of the ordinary, too. She certainly made her share of appearances on Lanesville TV, prompting the predictable isn't-she-cute calls. But her biggest contribution to the station was a piece of hardware that became a regular feature of the Saturday shows.

Sarah had the good manners to delay her entrance until after an interminable drive down the mountain and across the Hudson to the hospital in Rhinebeck. At each turn, Carol looked up from her Lamaze breathing to point me away from tortuous, back road short cuts. The night before she went into labor, we'd gone to the movies. As we returned to the house, I heard screams coming from the viewing room. Carol went straight upstairs, but I lingered in the doorway. Tobe Carey of the True Light Beavers had just returned from Mexico, where he had taped an American woman giving birth while lying in a hammock. The tape graphically depicted her painful and prolonged delivery. A dozen people had gathered to watch the playback, all of them enthralled by the drama. Video recordings of births are commonplace now, but in 1972, none of us had seen anything other than sanitized Hollywood depictions of deliveries–none of the anatomy, none of the fluids. They urged me to join them, offering to rewind the tape so I could experience the whole thing. I declined and followed Carol upstairs, a little shaken.

Carol had a much easier time than the woman on Tobe's documentary; easier, anyway, from my vantage point as an observer. We decided in advance to forget about taping. I could picture myself engrossed in some portapak malfunction just as Carol needed me most. Like our wedding, our memories of Sarah's birth and that of her two sisters are no less vivid for living only in our minds, unrevised by endless replays.

Sarah's daytime crib was her baby carriage, a classic perambulator, black fabric, metal frame, with a fold-down sun hood. It was handy around the house, but not very helpful otherwise. The driveway was bumpy, the shoulders of the highway were pitted and rocky, cars drove too fast; and there was no particular destination nearby that beckoned casual strollers. When Sarah outgrew the buggy, we stuck it in the second kitchen next to the washing machine and dryer and our equipment storage cabinet. The second kitchen, a helpful feature for a kosher boarding house, served us as a staging area. It's where we assembled our portapak systems and tested them before going out on a shoot. The carriage became a nuisance, in the way and useless until one day Bart, exasperated, plopped a portapak into it because there was no other surface available. A lightbulb went on, and he decided to take the portapak for a stroll.

He and his "baby" headed down Route 214. Two women we'd never met nor even seen before emerged from a small home three doors away from the Farm. They approached the carriage, hoping to see the baby they'd watched on TV. Bart had unmasked our first closet viewers. They agreed on the spot to an exclusive interview on Lanesville TV. Within a few more years practically every commercial TV station in the country would have its own "Action News" remote van, racing from crime scene to disaster and back again to bring us all a whole lot more up-close-and-personal with these events than we might ever want to be. Lanesville lacked frequent catastrophes to cover. But our mobile unit was out there long before the others, bringing our viewers eyewitness reports of what games the Benjamin kids were playing or who was raking the yard–and sometimes coming back with the news that not a soul was stirring in town today. The seeds of America's addiction to action news van coverage were sewn by the Lanesville TV Newsbuggy.

∞∞∞∞∞∞∞

The way we applied technology reinforced our aversion to hierarchy. Anybody could grab a portapak, like Bart with the Newsbuggy or Skip when he jumped in the car with a couple of Benjamin brothers and raced down to Chichester to cover a fire. Scotty talks tough on the way there–he can fight a fire, he knows how–until they near the fire engines and he melts back into a frightened five-year-old. The tape ends with the fire out and Scotty hitching up his pants, explaining how easy it was. We had hundreds of sequences in the same vein, solo projects that took no more than a portapak and perhaps a few hours in the editing room. The judgment of what to tape, how to tape it and when to show the result resided exclusively with the person who shot it. That type of autonomy simply didn't leave room for much formal structure.

On the other hand, Lanesville TV couldn't go on the air with fewer than two people, and a broadcast usually involved five or more. And whether for Lanesville TV or for our individual grants–the two were often indistinguishable–we all had projects that required assistance. We couldn't afford to be too choosy about our helpers, either. With only nine people and sundry guests on the weekends, finding a crew for the bigger projects, such as the Farkel-Van fight, usually came down to rounding up the requisite number of warm bodies. Davidson was the only truly methodical person among us, a trait reinforced by his training as a skilled woodworker. The rest of us prized spontaneity so highly, it took us years to appreciate the value of advance planning. That becomes painfully evident watching what's left of the tapes. What if we'd worked on blocking, lighting, sound and set design for more than a few minutes before airtime? Would anyone in Lanesville have noticed? Would anyone have cared? We constantly wrestled with our ambivalence toward production values, each of us pursuing a style of his or her own that wouldn't cross the line into the slickness, God forbid, of commercial TV. Each tape and each program became a new experiment. We had no conventions for content or style, no rules that couldn't be broken. We had our individual ways of working, and we preferred to rely on certain Videofreex rather than others at different times and for different reasons.

I asked Nancy to shoot *The Bunny Returns*, the Easter egg tape. She held the camera steady and had a good eye for framing action. Except for studio shows, we seldom used tripods. Handholding cameras is now standard procedure in the news business, but it was considered unprofessional when independent video and film makers pioneered it. Skip could have shot the bunny tape, too. And though Bart often used the camera in a style intentionally less fluid, he would have recorded equally usable tape. Neither David nor Davidson resided at the Farm at the time of my bunny tape, and Chuck's jitterbug camerawork reflected the abundance of nicotine and caffeine constantly in his bloodstream, as well as his inexperience behind the camera rather than inside it. I chose Nancy because I knew she wouldn't give me a rough time about making the tape her way. Working with her meant an effortless collaboration. She had the ability to separate her own projects, over which she exercised complete artistic control, from the work of other people.

In the best of circumstances, we offered each other not only mutual support but inspiration. And yet the very convenience of the group–the ready availability of skilled people with creative ideas–threatened us with staleness. We were a limited talent pool, and while we concentrated on working together, the video universe beyond the boundaries of Lanesville was rapidly expanding. By 1973, requests for information had increased to such a volume we couldn't keep up with it. Letters wanting to know more about us flowed in from Milwaukee, San Jose, Tacoma, Madison, little towns and cities like Fairplay, Colorado; Orcutt, California; and Louisa, Virginia, as well as from Germany, Belgium, Denmark and France. Had we relied solely on our own devices, we could never have kept up. But as the video spa of the Catskills, we could depend on a steady supply of visitors to keep us on our toes. And more and more, what the visitors came to see was not just the Freex, the video collective, but Lanesville TV, the experiment in community programming.

∞∞∞∞∞∞

In our efforts to build a local audience for the station we did what we could to overcome the prevailing belief among Lanesville residents that they really didn't have much to say to each other. Their collective reticence stemmed in part from traditions of self-sufficiency bolstered by isolation and a lack of strong community institutions. But just as much, it was rooted in old divisions, some of them along family lines–whispers of who fathered which children, of suicide, a child killed by his mother, of unyielding mistrust that defied mediation. Where we succeeded in breaching this wall of silence was with guerrilla raids led by the Newsbuggy and the Lanesville Players.

On a Thursday afternoon in the middle of August 1975, a group of us sat on the sun porch talking idly about topics for the show that week. I wondered out loud what would happen if we presented the people of Lanesville with a fictional situation that required a reaction. What if, for instance, an oil sheik wanted to buy the place? How would they respond?

The oil sheik idea came right out of the news. In the midst of the 1973 Yom Kippur War in the Middle East, Arab petroleum producing countries had imposed an oil boycott on the U.S. in retaliation for our support of Israel. The oil buying panic that ensued produced disruptions for the public and whopping profits for the major oil companies. There was gas rationing for a while, and the service stations in Phoenicia and Hunter posted "No Gas" signs some days, as supplies became erratic. We did a little less traveling, we car-pooled when we could, and like most folks, we managed to get by until the peak of the crisis passed and deliveries returned to normal. But by then, the price of a gallon of gas had risen from about forty cents to a dollar or more.

That presented a real hardship for all of us in Lanesville, because we had to drive everywhere we went. The Videofreex, luckily, did a lot of our business at home. But for most Lanesville residents, the cost of a trip to work, to shop, to the doctor or the movies rose dramatically during the mid-1970s. The whole country felt the same pinch, but the people of Lanesville took more of a hit because many of them lived close to the poverty line. Exxon could unapologetically make a 59 percent profit in the fourth quarter of 1973, because TV shaped the anger and mistrust of the nation's viewers, and TV depended on oil companies as sponsors. Sure, the public mistrusted oil companies, but TV had made sure the thrust of public hostility was aimed at greedy Arabs and not greedy oil executives or the thoughtless consumption that got us into dependence on foreign oil in the first place. The people of Lanesville, no matter how fuzzy their TV reception, had gotten the networks' message as clearly as any Americans. They knew what an oil sheik looked like and assumed that Arabs as a group had become rich and acquisitive.

Shamelessly playing off this stereotype, we came up with costumes, and the next day, we showed up at the post office as a sheik and his entourage.

We would not have fooled anybody had we wanted to. We drove up in Willie's Cadillac with Willie at the wheel, and neither Bart, playing the part of the Sheik in an authentic burnoose, nor I, as his military advisor decked out in my father's World War II tunic, looked like anyone other than the Videofreex. But when we walked up to Charlie Benjamin, Sr., working on a car in the parking lot, he doesn't bat an eye. I snap open a briefcase full of dollar bills and Bart offers to buy the place. Charlie politely refuses. He also declines to sell his children, despite a generous offer from Bart. "The kids aren't f'sale f'love nor money," says Charlie, calmly wiping the grease off a wrench.

We cross the road to Joe Keley's. "Sell? Sure I sell. Forty t'ousand dollars! ...", at which point he works himself into a tirade, condemning the people who won't buy his land and all his deadbeat tenants. Life invades art. He knows tape is rolling, and it only encourages him to rant.

We fled the Keleys', fearing we'd end up with little on the tape but Joe's convoluted sales pitch. We'd go a long way for improvisation, but not so far as allowing Joe to upstage us.

I hoked up an ending about oil discovered in Lanesville; we shot the last few scenes that afternoon and edited the show that evening, getting *The Sheik Who Shook Lanesville* ready for the show next day. At that point, we had to determine what to call all the people from the hamlet who participated. We could have identified each of them by name, but because we seldom took more than a verbal credit ourselves for Channel 3 shows, that seemed like too much trouble, especially because our Lanesville TV viewers knew all the people who appeared on the tape. We certainly couldn't lump them in with Videofreex; they would have resented the presumption of having any official connection with the Videos, which was how we settled on crediting our actor-neighbors with the collective appellation of the Lanesville Players.

∞∞∞∞∞∞∞

After three years of regular weekly broadcasts, with only a few weeks off and very few reruns, the novelty of the station had diminished and with it, the enthusiasm of the audience. We could count on Sam and Miriam watching, and Willie, too, and usually Harriet and some of her family. We were always thrilled when a previously unknown viewer called, even if the person registered a complaint. It meant someone had tuned us in, and that gave us a little jolt. *The Sheik Who Shook Lanesville* prompted our regulars to call and a couple of others, too, so we considered it a success. More important to us, we had moved from experiments with found events like the prizefight to a new form of intentional community improvisation. The concept had worked, which meant we could plan other appearances by the Lanesville Players.

The last and most elaborate production to involve the Lanesville Players was the UFO tape. By that time, we had established the genre and Lanesville residents, regardless of their feelings toward each other, were eager to cooperate with us. For the UFO tape, this included multiple takes requiring kids and adults to point toward an imaginary UFO. The alien space craft was added in post production. It consisted of a prosthetic breast suspended on a thread. It wobbled across the screen in the final cut, looking no less convincing than the flying saucers that graced many a B movie.

The UFO show relied on the talents of several independent video producers, including Tom Weinberg from Chicago, and John Keeler and Ruth Rotko from Woodstock, and marked the final high energy point of the work of Media Bus and Videofreex at Maple Tree Farm. Lanesville TV continued on the air for another year or more, but the UFO tape served as a coda to the broadcasts. Our interest in the station gradually petered out after that. We all had other projects we wanted to pursue, and Channel 3 had become a burden rather than a rallying point. The Farm remained an important institution in the video movement, and visitors continued to arrive for another year. But the small scale of Lanesville and our station, the things that had made our experiment work, no longer fit our larger ambitions, nor did either the community or the channel promise any way for us to pay the bills.

16
Elsewhere
Area

In late October 1972, we conducted an informal Lanesville TV presidential preference poll on Main Street in Phoenicia. The response pretty much predicted the national outcome: 60 percent favored Richard Nixon for reelection, 26 percent preferred George McGovern and 14 percent hadn't made up their minds. We didn't dwell on the inevitable. We reported the results of our poll on our October 22nd show and then segued to our first *Lanesville Country Kitchen*, with Bart announcing our intention to tickle the "parrots" of our viewers. Four days later, Secretary of State Henry Kissinger told the world, "Peace is at hand."

In December of '72 we gathered around a makeshift banquet table that filled the music room for a Freex Christmas feast. We were joined by Carol's parents and mine and other assorted relatives and friends. As we ate, United States warplanes were in the midst of a two week campaign of terror-bombing North Vietnam. There was no major rally for us to cover that December and no special call for anti-war tapes we'd already shot. A month later, January 27, 1973, the administration signed a truce ending America's combat role in Vietnam. The central event of our adult lifetime, the one overriding political issue that defined who you were and what you believed about the world, had officially ended.

Skip and I sent a letter through a contact in Paris to a North Vietnamese official, asking permission to travel to North Vietnam. We described our idea for a TV program that would educate Americans about the effects of the war and show the everyday lives of Vietnamese. We said we hoped such a show would be a step toward reconciliation between our two peoples now that direct hostilities between our governments had ended. We never received a response to the letter, and after a few months of waiting we shelved the idea without pursuing other avenues of access. Like most of our contemporaries and much of the nation, we turned our attention away from the war in Southeast Asia. Vietnam was over for us in terms of chronicling opponents of the struggle there. We had no way of knowing that just as the war ended the FBI opened its file on Media Bus. But one frightening unmistakable reminder of the war's violence did keep cropping up.

I had gone up to our bedroom early one afternoon in the summer of '73 to tuck Sarah in for a nap. The day was still. Far off, I heard a low, grating whine. It grew in volume over two, maybe three seconds until it exploded overhead with the force of thunderclap. A shadow passed over the house in a blink and disappeared. A dull roar echoed across the valley a few seconds longer. Trees rustled in its wake. Sarah wailed as I tried to comfort her. Chuck ran upstairs. Had I heard it? I cradled Sarah in my arms and motioned for him to leave so I could get her back to sleep.

The flyovers continued every few weeks, always in the daytime, but never predictably. We could catch a glimpse of silver wings and fuselage always headed southwest down the valley. But they moved so fast we couldn't identify the shape or marking of the jets before they disappeared beyond the next fold in the mountains. Sometimes they came in pairs, all the more terrifying. I tried to imagine these same howling monsters dropping cluster bombs, napalming Lanesville. It didn't register.

I checked with local authorities and then the Federal Aviation Administration. No one admitted knowing anything, and the flights continued.

I was spending whole days on the mountainside late that year in the vain struggle to protect our receiving antenna from voracious porcupines. Early one afternoon as I worked at the antenna site, the plane flew by again. The bare trees gave me a commanding view of the valley below. For the first time I could see the jet clearly, its overall shape and the strange protrusion rising immediately in front of the cockpit. I was looking down on it from above. I rushed back to the house and sketched an outline of what I'd seen, which I mailed to the FAA, requesting identification.

The plane was a Grumman A6-E, an all weather attack bomber designed to be launched from an aircraft carrier. The distinctive proboscis was a mid-air refueling tube. The A6-E and other Grumman planes were manufactured by the military contractor on Long Island. I suspect company officials were caught off guard by my drawing. It would be hard to deny that their jets were flying at dangerously low levels down the Stony Clove valley after I described the top of the plane rather than the bottom. Letters went back and forth for a few months. I wasn't the only person to complain. Grumman officials asked for a special meeting with the Town Board.

People packed the town hall on the evening of the meeting. Grumman sent three representatives: a company executive, a test pilot and a Navy commander dressed in summertime duck whites, with gold braid on the visor of his cap. They flew into Kingston Airport in a plane so big it nearly overshot the narrow, lumpy runway. Town police cars chauffeured them to the mountaintop. The commander presented John Glennon with a model of the plane and a number of other ceremonial gifts for the town, as if he represented a foreign government on a mission to a local potentate. Then he asked to have the lights lowered. He pulled up a screen and the test pilot, after some embarrassed fuss, started a projector loaded with a Navy film on the virtues of the A6-E.

The plane carried a Hughes Aircraft radar terrain mock-up guidance system. It provided a schematic representation of the terrain ahead of the aircraft, a virtual reality of sorts, which, according to the enthusiastic narrator of the film, allowed pilots to navigate no matter what the weather. The planes in the film dumped their bombloads precisely on target and flew off into the sunset. The lights went up again and the officer explained that the Navy required Grumman to test each plane's guidance system before the aircraft entered active duty. The test pilots flying at hundreds of miles per hour above the treetops of Lanesville were watching little green screens with white lines, checking for flaws in the blind-flying system.

We were taping the presentation for Lanesville TV, and by the time the film ended, I began to feel uneasy about the outcome. People nodded at the film. They craned their necks for a good look at the model. The commander presented a dashing figure in his starched uniform. Pride in America swelled. Gordon Gruenwald, Fran's son and no friend of Videofreex, said he'd worked for Grumman, a real good company. No one should want to make trouble for a firm that helped defend the country. He wasn't looking at me but his message about our lack of patriotism was clear.

A local pilot raised his hand and waited patiently for Glennon to recognize him. He flew out of Maben Field, he said, a grass strip above the nearby hamlet of Lexington. Didn't the jet flights pose a danger to private fliers, he asked, especially if no one knew when the planes would swoop into this region?

Well, smiled the commander, they only flew on weekdays, and most private fliers went up on weekends. And you never find much air traffic up here, anyway, because this is classified as "an elsewhere area." People shifted in their seats. I could hear grumbling. The Town of Hunter, the center of the High Peaks of the Catskills a what? An elsewhere area? Hands rose on all sides of the room. What did he mean by that? Did elsewhere area mean Grumman could do anything it wanted here? Why not buzz the Adirondacks? The mood in the room had suddenly changed from pride to anger. Our neighbors began speaking up about the disruptions in their lives caused by the noise of the flights. Tempers rose. Glennon looked at the commander. Time to close the meeting. But before he could speak, Town Justice David Jacobs stood up. As a former board member and the owner of a successful electrical contracting business, his word carried a lot of weight. He lived on the other side of the notch and I figured the planes didn't come nearly as low over his home as they did over ours. He understood the necessity for the flights, he said quietly. Then he paused. But when the jets came over his house, the dishes in the kitchen rattled and the chickens wouldn't lay for a week! He boomed this last part. It ended the meeting. For the most part, it also ended the flights.

∞∞∞∞∞∞

Skepticism about the Paris truce tempered our desire to express relief. The federal government had lied to the country before, and a temporary peace in Southeast Asia augured no major changes here at home. After all, Nixon had just taken the oath of office for a second term. Four more years of him seemed like an eternity.

Opposition to U.S. involvement in Vietnam had been the one topic on which we could all agree no matter how disparate our views on life, art, video or the prospects for a just society. With the bombing and killing, and the draft and the marches a matter of history, we would have to choose our causes rather than have them pick us, the way the war had. And the roster of choices grew longer each day.

Nixon could have become the issue, except that not long after the peace treaty, he found himself in more hot water than we could have hoped to make for him. On August 8, 1974, the Videofreex gathered in the viewing room with Sam, and Earl and Maria Voorhes to watch the president's resignation speech. Skip had left for Washington, D.C., a few days earlier to tape reactions outside the White House for TVTV. Later that week he gave the Channel 3 audience his impressions of the nation's capital at the transition. He couldn't add much to what the networks had already said, but hearing it from him, a neighbor, offered the Lanesville audience a sense of connection to a major national event.

It took nearly another year for America's surrogate war in Vietnam to end. Saigon fell in April 1975, as did all of Cambodia not already in communist hands. The war, which had taken a million or more lives, over 50,000 of them Americans, had culminated in the very result every president from Truman to Ford had tried to prevent. We could only marvel at the waste of lives, of time, energy and resources; and we could hope Vietnam would move on to a peaceful future and so would we.

But the war at home continued. The post-Hoover, cleaned-up, non-political FBI considered it necessary to keep spying on us for two more years. Force of habit, I guess. Loose ends, what with the Weather Underground people still on the loose. Paranoia dies hard, especially in a world of secrets. And what if we harbored fugitive heiress Patty Hearst? We didn't, but the domestic spying apparatus couldn't let go. And if they'd watched Lanesville TV the night Michael Shamberg showed up with his tape of Abbie in hiding (we have no information they did watch or that any of their local snoops reported it), that certainly would have fueled their fears.

Abbie was busted in a cocaine deal in 1974. He claimed he was only an observer, a journalist. Somebody set him up, he said. It sounded lame, but with Abbie, you never knew. Rather than face a long prison term, he jumped bail and went underground. Completely disappeared. I half expected him to show up at the farm one day.

After Abbie became a fugitive, Michael put out the word he wanted to interview him. It didn't take long before he received a response from some people Glennon would once have been very keen to meet. They said a tape with Abbie could be arranged under certain conditions. One of those conditions was money. Michael describes what happened next in an interview conducted by Megan Williams, another of TVTV's founding members and Michael's wife. Her interview with Michael is woven into the interview with Abbie. Only Michael, Ron Rosenbaum, a writer on assignment for *New Times* magazine, and a few companions inside TVTV knew how close they were to connecting with Abbie; Michael was ready to go when the call came. He was told to show up at a parking lot in California. Strangers met him there and gave him opaque, wrap-around sunglasses. After a long drive they led him to a room, he doesn't know where. Abbie was waiting.

Abbie and Michael knew each other from TVTV's *Lord of the Universe*. Abbie had provided a running commentary on the teenage guru and his uncritical young followers, including the most memorable line of the program: If this kid is God he's the God that the United States of America deserves.

As the interview proceeds, Abbie can barely control his hostility toward Michael. Their egos and styles could only have clashed. But neither of them was willing to pass up the opportunity, Abbie to tweak his pursuers, and Michael to claim a major media coup. They spar without much to show for it. Abbie lectures. His twisted sense of humor, the quality that elevated him above the status of a politically hip wise guy with a bad attitude toward authority, appears in flashes but never engages. He chooses his words too carefully. The strain of living on the run shows through. It's painfully clear that just outside the door, someone is listening.

Michael had to leave the tapes for two weeks with Abbie's underground handlers so they could review what had been recorded. When he got them back, he put together an edit and brought the final version to the farm. I never saw him more excited and more pleased with himself. We premiered the tape of Abbie as a fugitive on a regular Lanesville TV show in the spring of 1975 and got no reaction at all from our viewers. Michael didn't let on if he was disappointed. He left the next day for New York City to go after what Abbie really wanted–the one thing the government assumed it had deprived him of once he became a fugitive: a mass audience.

TVTV had a deal to produce a number of documentaries for The Television Laboratory, a project of public television station WNET, Channel 13 in New York City. With funds from the Rockefeller Foundation, David Loxton, the English director of The TV Lab, set up shop in a basement studio near the United Nations on the East Side of Manhattan, not exactly under the thumb of WNET station executives across town, but still well within the gravitational field of Midtown media power. Loxton had a commitment from the station for a Sunday night show, Video/Television Review, or VTR, the only regular broadcast of independent video in a major market. One of the early VTR shows was called *The World's Smallest TV Studio*, an edited version of a Lanesville TV broadcast showcasing some of our best work. But TVTV was a far more regular contributor to Channel 13 than Videofreex.

Critical acclaim for the '72 convention tapes had opened the door for *Lord of the Universe*, which eventually won TVTV a duPont Award, one of the most prestigious in journalism. The group went on, still with a number of the Freex involved, to produce *Adland* and the *Gerald Ford's America* series in 1975. TVTV was hot and Loxton's TV Lab was heavily involved in the funding the group's tapes.

The tape of Abbie, called *In Hiding: An Interview with Abbie Hoffman*, was no surprise to Loxton or the station. TVTV not only proposed it in writing, the proposal specifically stated, "TVTV has potential leads to Mr. Hoffman." The tape of Abbie got a better time slot than the VTR show, too, because a Scandinavian documentary was pulled at the last minute after critics called it racist. On May 19, 1975, the day of the broadcast, a two-column ad appeared in *The New York Times* listing the nine o'clock showtime, with a big picture of Abbie and copy that touted the "exclusive" nature of the interview and "the cloak-and-dagger arrangements that made the interview possible." In an unintentional irony, the ad ran next to a story of the opening of the civil trial against the governor of Ohio and 52 members of the Ohio National Guard for gunning down unarmed student antiwar protesters at Kent State University exactly five years earlier.

The tape of Abbie might have passed from notice as a curious but unexceptional journalistic exploit but for another story in the May 19 edition of the *Times*, a review of *In Hiding* by John J. O'Connor, the paper's TV critic. O'Connor explained toward the beginning of his piece that Abbie had insisted on being paid $5,000 for the interview, and that Shamberg responded by paying him $2,500 and turning over a video cassette player reportedly worth $800. Rosenbaum paid an additional $500.

Two days later, *Times* television reporter Les Brown ran a piece about how both Channel 13 and Fred Friendly, a consultant for the Ford Foundation, had "denounced" the program. Friendly, a former head of CBS News, demanded reassurances the foundation's money hadn't been used in yet another case of what had recently come to be known as "checkbook journalism." That term was coined after CBS paid former Nixon aide H.R. Haldeman $50,000 to tell his story to *60 Minutes*. Friendly fulminated about how the "goodwill and good name" of the Ford Foundation had been tarnished by the credit the Foundation received at the end of *In Hiding*. That made it look as if the Ford Foundation had not only doled out cash for the tape, but had actually paid a wanted man to deliver a biting condemnation of the American political system in prime time throughout the nation's largest TV market. Friendly wasn't placated by assurances that TVTV had forked over its "discretionary funds" to Abbie.

Faced with the ire of a major funding organization, the station at first took the easy way out: it pretended ignorance. George Page, the Channel 13 executive with responsibility for the TV Lab, assured the *Times*, "... We're quite upset about it [the payment] and wouldn't have allowed it if we knew." But a few days later, the real story emerged when Brown interviewed Gregory J. Ricca, a lawyer for the station, who admitted WNET "conducted legal research before the broadcast to establish that there were no criminal implications in making contact with a fugitive." (We could have told them that based on CBS's decision to air *Eldridge Cleaver*, but nobody asked us.) Ricca went on to tell Brown, "... We were also satisfied that we were not aiding and abetting a criminal, even with the payment of money." On May 24, the day the interview with Ricca appeared, Channel 13 rebroadcast Michael's tape of Abbie.

I remember one other discussion of *In Hiding*, which occurred when Michael arrived at Maple Tree Farm with the tape. He thought any potential liability faced by Channel 13 for consorting with a fugitive might be diminished by playing the tape first on Lanesville TV. That way, WNET/13 was only repeating a program already shown elsewhere. He didn't think it mattered that elsewhere in this instance meant our illegal TV station. I hesitated. We could find ourselves holding the bag if federal authorities challenged WNET's right to broadcast the tape. I didn't like the idea of protecting a hidebound institution like Channel 13. And it was clear the station would have no interest in defending us if some sort of court battle ensued. Despite these reservations, I couldn't pass up the opportunity to play the tape, nor could the other Freex, so we agreed it should air on Channel 3 before the New York City audience saw it, whatever the consequences.

The question of who first broadcast the tape was never raised by the *Times* or in any other public forum. Friendly remained grumpy about the show, calling it a "rip-off." Channel 13 President John Jay Iselin, after letting his subordinate make a fool of himself, said later he supported Michael. And Michael offered a rambling justification for the payment, saying that while TVTV didn't believe in paying for a story, this case was different, and besides, Abbie had to use some of the money for "defraying his expenses in connection with the show." We experienced a little nervous excitement in anticipation of running the tape in Lanesville, but once it went on, we knew we'd done the right thing by helping Abbie get more uncensored time on TV than he'd ever had before.

∞∞∞∞∞∞

The only time I saw Abbie after we moved to Lanesville was the flight to John Lennon's birthday Party in Syracuse. As we waited to board, he and Jerry Rubin were giggling about finally having the chance to ride in a first-class compartment. They looked disappointed when they saw the chartered jet had only open seating front to back. Other than Michael's tape, we knew nothing of Abbie's whereabouts from the time of his bust until he turned himself in to authorities in September 1980.

I was living in Washington, D.C., at the time, and had just written to Don West, my first attempt to contact him in a decade. After Don was fired from CBS, he'd kicked around the periphery of the TV business for a while before landing a good job as editor of *Broadcasting* magazine, the voice of the commercial broadcast industry. I wasn't sure whether he'd respond, but he wrote back to say what a surprise it had been to receive my letter the same day Abbie surrendered. It turns out Abbie had been living a public life as an environmental activist named Barry Freed in the island community of Fineview, N.Y., on the St. Lawrence River, just down the street from Don's summer home.

Abbie served a short sentence in a prison/halfway house and later moved to the hinterlands between New York and Philadelphia. A documentary film on him made before his death captures flashes of the old wit and insight breaking through a heavy veil of what looks like fatigue. Friends said later it was severe depression. In the film, it's as if he's searching for a way to make himself relevant, for a path to regain the following that fed his manic energy. He might have found it, but the disease silenced him first in a way the government couldn't. He killed himself in 1989 with an overdose of alcohol and drugs.

Before his death, Abbie described his take on what had happened between him and Videofreex. The passage appears in his last book, *The Best of Abbie Hoffman*:

One day while turning on with the Video Freex, a New York media collective, I asked if it was possible to pirate an image onto network television. Curiosity flickered faster than a strobe light. Equipment was bought, tests made, and one evening while David Brinkley was analyzing the news, a couple fucking appeared on a number of sets in the Soho area of downtown Manhattan. Eureka! It worked. The Video Freex freaked, gathered up all the equipment, and hid out. "But we saw the fingers on the monster move," I pleaded. It was no use; their equipment and license were more precious than jewels. The technique lived on, however, in the pages of Steal This Book.

Ah, Abbie. You always knew a good story would outlive the facts. I don't doubt for a minute you could see "the fingers on the monster move." You always saw things more clearly than the rest of us. But we didn't think of it as a monster. It was only a tool. And if you thought your monster drove us into hiding, no wonder life on the lam never brought you through Lanesville. Too bad, too. You would have made a great Lanesville TV host.

17
You Fallin' Apart Here?

Nelson Rockefeller was named vice president by Gerald Ford in 1974, several months after he voluntarily resigned as governor. His departure from the governor's mansion in Albany after fifteen years in power didn't mark the end of state support for video, but the funding priorities of his beloved Council on the Arts soon came under review. Conservative legislators never tired of complaining about allocations for the arts, and a state senator called in reporters one budget season to single out Media Bus as an perfect example of how government was wasting taxpayers' money. Imagine, a media center (with the emphasis on "center") in a little place like Lanesville. This type of external pressure, compounded by conflicts within the group, led most of us at Maple Tree Farm to get serious about finding other work.

The beginning of 1975 found me at a casino on the Belgian coast. Wendy Clarke, Shirley's daughter and a video artist in her own right, asked me to assist her in setting up a multi-monitor video exhibit at an international festival of experimental film and video held every five years at an ornate casino in Knokke, a North Sea resort near the Dutch border. It meant a free trip to Europe, and I took it. Film remained the prestige medium on the Continent, and film makers and critics could hardly be troubled to grace the video exhibits with their presence. But young Europeans swarmed around the Americans, who dominated the video section of the casino, earnestly questioning us about everything we did.

Carol and Sarah met me in Paris when the exhibit ended, and we stayed once again with Carole and Paul. They had fallen out with Jack Moore, and that made it uncomfortable for us to try to connect with the "Pope of Video" and his international video scene. Jack was lording it over everybody from his Paris headquarters in the atelier of a partner of his, a place crawling with sweet-faced young men. He knew a Sony executive who, for unknown reasons, entrusted him with keys to the Sony showroom on the Champs Elysees. He and his crew could use all the Sony hardware when the showroom was closed; they could place long-distance phone calls and, most important, transfer tapes between European and American video standards. Jack bubbled with stories. He'd had a big contract to project experimental tapes at the Munich Olympics in 1972. When terrorists took Israeli athletes hostage, officials excluded Jack and his crew from covering the tragedy as it unfolded. "Well, of course," said Jack, dry and prissy, "we didn't make any friends when I pointed out 11 Jews is no record for Munich."

Comments like that didn't boost his popularity among our German video friends, either. Many of them lived in Munich, and one, Charlie Roesch, was in Paris. Charlie–his name was Karlheinz, and he was taking flak from German friends for adopting an Americanized nickname–drove us to Germany to the headquarters of the West German TV network ZDF, to preview a documentary about Videofreex and other groups he and his friends had shot in Lanesville and New York the previous year. In the middle of the playback, the monitor went dead. After some terse words through the intercom, our network hosts apologetically reported a technician in the basement had turned off the VTR. It had not been properly scheduled. Achtung!

Judging from the number cities in Europe where we could find a video group with floor space for us to sleep, the movement was gaining strength internationally. A European standard had emerged for half-inch VTRs, and we ran into a lot more people with portapaks than we had three years earlier. The technical barriers had fallen, yet stubborn cultural prejudices persisted. Other than Jack's nomadic video caravans, we didn't find much enthusiasm among video people for exchanging tapes within Europe. German and French video artists, like their American counterparts, were looking inward, except where the U.S. was concerned.

West Germany was the only major country where government controlled television networks didn't ignore young video producers. The show made by Charlie and his friends played in prime time on the national network. There I was with my halting German, on TV all over the country. By contrast, Carole and Paul were still struggling as outsiders in Paris. French political lines were rigid and the two of them didn't fit in. To give us a sense of this, they suggested we visit the massive headquarters of the French Communist Party in Paris. The party, I gathered, didn't approve of video. It made films, hours and hours of color films backed by an extensive, nationwide distribution system. We were directed to take seats in the viewing room, only to find ourselves trapped in front of the single most deadly boring movie I have ever seen–a three-hour extravaganza on a peaceful strike by workers at a watch factory. Desperate for a way to escape without offending our hosts, we pleaded for the lights after discovering Sarah had pooped in her boot. You could see by the disdain on the faces of those dedicated party members they regarded us as hopelessly frivolous American TV dilettantes. By their standards, I'm sure we were.

In late 1973, Carol and I had received a letter from Eldridge Cleaver. After a long struggle to win political asylum in France, he had left Algeria, abandoning tapes and video equipment. He was living in Paris and was eager to learn what was happening with video in the U.S. He wanted to develop contacts through Carole and Paul, and asked that we write back to him, addressing our letters to "Andy" at the Rue d' la Odeon apartment. A short while later he called the Farm and spoke to Carol, repeating his request for information. I wrote him a long letter, summing up the video scene in the U.S. We received no reply.

In Paris a little more than a year later, Carole asked if I wanted to meet Cleaver. Is he still into video? I asked.

She had to think for a moment. I sensed they weren't on good terms and she was making the offer because she felt I expected it. No, she said. He's not very political right now. No video. We don't see him very often.

I couldn't think of what I'd have to say to him if we didn't talk about video, so I declined the invitation. Carole shrugged and didn't bring it up again.

That summer Cleaver told *Newsweek* he wanted to return to the U.S. In his absence and in the wake of the Watergate scandals, a steady stream of information had emerged documenting the actions the FBI had taken to attack and destroy the Black Panthers. Attorney General William Saxbe acknowledged the Bureau had conducted a systematic program of harassment for four years, and his revelations did not end the matter. Cleaver returned to the U.S. in November 1975 and was released on bail after telling the judge that during his exile he had undergone a "spiritual conversion" and become a devout Christian. I heard later he had introduced a line of men's clothing, which featured pants with codpieces, a throwback to the medieval practice of placing a cup of fabric over male genitals. I saw him a few years ago on a religious TV station. Like Abbie, he sounded tired and subdued. I surfed on without listening to much of what he said.

∞∞∞∞∞∞

The trip to Europe in '75 was exhausting for Carol. She was pregnant again, and by the time we returned to the farm she had become seriously ill. We were afraid for a while she might lose the baby. But Emilia was born that summer, a healthy, happy infant. Now we had two kids, and Carol and I still earned a combined weekly salary of fifty bucks plus room and board.

Not only had the war and the draft and Nixon's reign as imperial president ended, Congress had undertaken an investigation of the abuses perpetrated by our spy services, making it appear as if the forces of repression in this country had sounded a retreat. Militants got jobs to support their families. The maw of the entertainment industry digested the remnants of a counter culture no longer able to find a target to counter, and out the other end came disco music and *The Brady Bunch*. Our status as rebels, which had impelled us to band together for mutual support, evaporated. For years, we trumpeted our lack of a group ideology as the key to our success in maintaining individual identities within the group. This cohered the collective, we said. Now we asked ourselves: What do we have in common?

Carol and Nancy could rely on a friendship that went back well before Videofreex. And in the group, they both learned early on that if they wanted to pursue their own projects, they would have to assert themselves: I'm sorry, but today I need the portapak ... the editing room ... the last tape on the shelf.

After CBS ushered Don to the door, Carol returned to teaching school in Harlem while Nancy took a job with ASCAP, listening to tapes of radio and TV shows, checking to make sure stations paid the proper amount in royalties. Their real world jobs made it that much tougher for them to stay involved in the group, but they persevered.

Videofreex men, on the other hand, behaved like macho knuckle-draggers, assuming our chicks would naturally take subservient roles, letting us make tapes while they typed and helped out when we needed them. That ended for good when we arrived at Lanesville with no master plan for how we would function. Each of us but Chuck had to define his or her job. Artist didn't count. And out of that chaos, we became an equal opportunity employer of good ideas, with the women producing just as many as the men. Curtis had the roughest time. After CBS, she decided to take a break from video and travel. When she returned to claim her place in the group just before we left the city, she arrived alone, temperamentally ill-equipped to fight for a position of her own. Her marriage to Cy shortly after we got to the farm offered her a ticket out and she took it.

When Ann arrived in New York as Chuck's girlfriend, no one knew her skills and none of us bothered to ask. He bugged her to get a job and an apartment so he could move in with her, as I had with Carol, and Bart with Nancy. But she couldn't find jobs better than the dreary clerical position she'd left at the Rose Art Museum at Brandeis. So she stuck around the loft, a presence but not really a contributor until someone discovered she could draw.

We immediately piled on requests for her to letter the credits for our tapes and to whip out illustrations for brochures. We didn't have a character generator at that time, a computer that produced typed words on the video screen. Whenever we used credits, and we often ignored them, we created the words on cards with press-on type, one letter at a time. Ann's drawings and hand lettering added class to our tapes. Her illustrations also brought the pages of my book to life. We worked closely for months, with me scratching out rough ideas and her turning them into recognizable images.

The book sold surprisingly well, thanks in part to her drawings. French and Dutch publishers bought the rights but never issued translations. A group of industrious Germans translated the text and modified it for European video standards. They sold it as a bootleg edition with no royalties for us. As pirate broadcasters, we could hardly complain.

My parents looked forward with pride to the publication of my book–their son, the college drop-out, an author. I could read the pain on their faces when I showed them the finished manuscript. My name appeared nowhere on it. I had decided from the outset I would attribute *The Spaghetti City Video Manual* to Videofreex, believing we should all promote the group ahead of our individual achievements. I made the choice on my own. Nobody pressed me to do it and nobody emulated me, either.

My decision to leave my name off the book was not entirely a selfless act. I worried about ending up type-cast as a techie. I viewed myself as a producer with technical knowledge, but I sensed that if this book circulated widely in the video world with my name on it, I'd never be taken seriously as anything other than a tweaker with a good vocabulary. So when it came time to prepare the author profile on the back cover, I wrote a cryptic line about us being "an innovative group concerned with uses of video," which accompanied a photo of all of us standing in the doorway of the American Legion hall in Hunter. I dedicated the book "To Chuck and to Ann ..." without mentioning their last names. Ann, however, had the presence of mind to sign her cover illustration.

In an interview taped by Nancy, Annie explains how she had to beg for something to do other than sit with the equipment while the rest of us shot tapes. Was this the Annie I practically had to drag into the editing room? It turned out she has an eye for editing. It's not a physical skill so much as a sense of timing and how things fit together visually. Annie, who never undertook a major video project of her own while she lived with Videofreex, became one of the most accomplished video editors in New York City after she left the Farm, much in demand by Barbara Walters and other TV luminaries.

There was one area over which Ann exercised indisputable authority at the farm–the garden. In neighboring Delaware County, an old saying has it that the land yields "two stones for every dirt." I'd call that a conservative estimate for Lanesville. In the small meadow that wrapped around behind the Harley and Lindsey houses next door, Ann staked out her plot. Skip and Chuck, among other Freex, provided stoop labor, but we always called it Annie's garden and properly so. She managed to coax amazing bounty out of that rock strewn parcel. It may have been the biggest garden in Lanesville. It certainly fed us well.

Ann and Chuck split up after a couple of years at the Farm: irreconcilable differences, but no outward bitterness. They both stayed on at Lanesville, living separate lives; and one year Annie decided to take a vacation in the early fall, just at harvest time. The garden produced too much of everything that season, especially squash and tomatoes. And in Annie's absence, none of us would volunteer to take on the task of canning, freezing or otherwise preserving the fruits of her labor. Heeding forecasts of a killer frost, we picked what we could and announced on Lanesville TV we had vegetables free for the taking. Anyone who wanted them should show up at the Farm the next day. The following morning people we'd never seen before overran the driveway. We didn't recognize any of them as folks who called the show. They swarmed around the baskets of produce and carted away shopping bags full of veggies. When Ann came home she was pissed to find the vegetables gone. No one apologized. We felt we'd had a public relations coup for the station: our first and only Lanesville TV giveaway.

When Videofreex first got to Lanesville, we had agreed during one of our meetings to post the sign above the stove suggesting jobs visitors could do to help. Maybe we should have invited more obsessive guests like Charlie and his German friends, who treated the sign as a list of commandments. Most of our visitors took advantage of the hotel service we provided and left us with varying degrees of mess to clean up. The custodial chores were becoming a burden. No one wanted a repeat of the stuck-together sheets, and yet most of us pleaded we had too much work to do to handle the serious cleaning and washing. So about the time Emilia arrived Carol proposed hiring someone to clean the public spaces.

Ann opposed the idea. Having a cleaning person would amount to an invasion of her privacy. She wouldn't hear of it. Carol stood her ground, and Ann laid down an ultimatum: If Carol insisted on an outside cleaner, she'd leave. Nancy tried unsuccessfully to mediate. Neither side would yield.

I went downstairs to the kitchen and sat at the table the day Ann announced her departure. Shades of Frances and me. Chuck, Skip and Bart circled, accusing me and Carol of breaking up what was left of the group. Nancy came into the kitchen but couldn't make herself stay. I kept my peace. The confrontation petered out. Nothing more to say.

I went upstairs and found Annie in her room. We talked quietly for a long time. The cleaning issue provided her cover; easier than saying, I've had it, and facing the guilt sure to erupt from the group. She could find work at the farm, but nothing satisfying. Our social life revolved around video and guests. That sufficed for a recluse like Chuck, maybe, or for the Freex who traveled with TVTV, or for Carol and me, with our family. But she wanted to get on with her life, and Lanesville offered nothing but dead ends. Her honesty prompted me to admit that Carol and I had decided to move out, too. We hadn't told anyone else. I asked her to keep our news to herself. She agreed and we remained friends.

I helped Annie move back to New York, to a tiny apartment on 8th Street with a window that opened onto the exhaust vent for a Famous Nathan's hot dog restaurant. The last item out of the car was her cactus, which left small needles under my skin, an irritant that lasted for days after I returned to the farm.

∞∞∞∞∞∞∞

Maple Tree Farm could have accommodated our growing family, especially with fewer Videofreex than ever. David had moved to Boston to temporarily take over his father's business. He stopped at the farm the summer Emilia was born to show off his new pick-up with a camper on the back, which he shared with Moses, his sweet-tempered, foul-smelling basset hound. He hit the road and seldom came by Lanesville again after that. Davidson had long since established himself in the city, and by 1975, he no longer returned even for hunting season. We added no one new to the group, either, except in a peripheral sort of way. Jane Aaron, an animator and a friend of Ann's, took the small room in the extension across the driveway from the kitchen. She worked there and in the city on her own films and on commissions from *Sesame Street*. Jane ended Skip's claim as the most eligible Videofreex bachelor, though she kept her space in the back and he kept his room on the third floor as their romance proceeded in what looked like fits and starts. We could count on Jane to jump in behind a camera if we needed her for Lanesville TV, and she drew imaginative backdrops for sets. But unlike her female predecessors, she found no incentive to merge her identity with the group's. Instead, she and Skip gradually disengaged from Videofreex.

That left as many rooms as we could have wanted for Carol, Sarah, Emilia and me. But rooms weren't the issue. We wanted to go to the kitchen and not have to tip-toe around discussions of video theory, a place where we didn't have to worry about tying up the bathroom when a kid needed to soak in the tub. We wanted our own home.

I had thought we should make a profound break, like Ann, David and Davidson, move to the city or some other state. Carol pointed out the obvious. I didn't have a job and if we moved, she wouldn't have one, either. So we found a run down Victorian in Phoenicia, and moved out of the farm that fall.

At first, it seemed as if we hadn't left at all. I went to work in Lanesville most days, and on Saturday evenings, we'd pack up the kids and help with the show. Sometimes the needs of the children didn't match the broadcast schedule. Carol would stay home. She'd ask me why I felt obligated to continue working on Lanesville TV when we didn't live in Lanesville. I didn't have a good answer.

∞∞∞∞∞∞

Channel 3 shut down for a while in the early summer of 1976, when we went to Goddard College for several weeks to teach at the summer media program there. Goddard, a radical experiment in four-year undergraduate, liberal arts education, was the site of one of our early triumphs and a near disaster. At the national Alternative Media Conference in the summer of 1970, we debuted the Blue Calzone, the world's first inflatable TV set at Goddard. It was Davidson's idea, a huge blue and pink inflatable structure with a eight-by-twelve foot rear projection screen forming one end. Except for the color and the tubular tail for the fan, it did look a little like a giant Italian pastry. Inside the Calzone, we placed our video projector and a VTR. When it got dark, we played tapes for people sitting outside. As the breeze blew, the set would drift ever so slightly from side to side, almost as if the screen resided in a gelatinous display.

The Blue Calzone was a popular exhibit on the lawn near the small dorms of the rural college, but the best thing about it for me was not watching tapes on the screen, but sleeping inside it during a thunderstorm, as lightening strobed the screen. It became a popular refuge in the storm, pronounced "very trippy" by those who would know. That was our success.

The near disaster was a sign Chuck posted on a bulletin board during the Alternative Media Conference announcing that the Videofreex had called for a "fuck-in" at a nearby pond, and that we would tape it. Feminists and other rational people found this offensive proposal nothing short of exploitation. The event was called off, apologies were offered and, warily, accepted.

We were happy to return to Goddard in 1976, knowing it was a loose atmosphere that encouraged students to experiment. It was the only place we ever tried to duplicate Lanesville TV. We experienced technical problems, but the students loved the idea and made original tapes for the Channel 3 clone. A few hundred miles west at Syracuse University, students we didn't know were making national news with a hit-and-run broadcast of pornographic movies to campus dorms. The stunt didn't catch on, but the episode made it clear to us and our students we had no monopoly on pirate TV technology.

In the middle of a particularly hot night at Goddard, I bolted upright in bed, startled by screams coming from a nearby dorm room. The screaming stopped and I heard calmer voices through the walls. I went back to sleep. The next morning Nancy told me she thought someone was attacking her in her sleep.

Between teaching assignments at Goddard we began to plot how we would cover the Democratic National Convention that summer in New York City. TVTV had expressed no interest in trying to duplicate its 1972 tapes. But a group of us, including Tom Weinberg and DeeDee Halleck, thought the convention was worth documenting in a very different way. Collectively named *The Five Day Bicycle Race*, our series of prime time shows on cable TV during the convention patterned themselves after Lanesville TV, with a mix of live segments and tape, and a phone so viewers could call in comments. Scores of producers we'd worked with at Lanesville participated in what amounted to the realization of the challenge I'd laid down to Michael four years earlier. Michael didn't join us. He had moved to California some years before, preparing to leave video behind for a career producing movies.

The Five Day Bicycle Race won critical acclaim as the only alternative view of the convention, and inspired DeeDee and me to try to duplicate that success with a live election night show in New York City. We asked video crews to roam the city and rush their tapes back a la the Lanesville TV Newsbuggy. That show, called *Mock Turtle Soup*, muddled its way through the evening, another example of the Lanesville TV style, made available via cable to the nation's largest television market.

Nancy and Bart didn't work on *Mock Turtle Soup*. At the time she imagined the attacker, Nancy noticed a lump in her abdomen. By the fall, tests showed a tumor. A local surgeon said he hadn't operated on a tumor like hers before, but he'd be glad to give it a try. She thanked him and chose Albany Medical Center, instead, where doctors removed the malignant growth. She recovered in a remarkably short time and experienced no recurrence, but the gravity of her illness threw all of us off balance.

In early 1977, Nancy went to Detroit to visit her mother, and we continued the shows while she was gone, agreeing that we would give ourselves a break when she got back. Shortly after she returned, I was setting up for a show when she walked by the control room door on her way upstairs. "Do you think we should announce that we're going off the air for a while?" I asked.

She stopped on the stairs and gave me a condescending smile. "You have to do what's right for yourself," she said sweetly, and continued up the stairs. A hostile response to a threatening question.

Nancy, Bart and Chuck, and Skip and Jane when they were home, continued the broadcasts with the help of weekend guests. They saw it as the reason why people came to Maple Tree Farm, the one activity sure to keep the farm vital. I figured the experiment had run its course, and why kick the carcass. Easy for me to say, since I didn't live there. I backed off and started to find other things to do Saturday evenings.

We still worked together as Media Bus, despite periods of unemployment caused by late payments from the Council and other funding agencies. Nancy and Bart traveled more frequently than ever, and Chuck would split when he felt like it, which occasionally left the house unoccupied, or seemingly so. Chuck's shop was now in Davidson's old third floor loft, and it was hard to tell when he was home. I was wandering through the deserted first floor one day when I came into the kitchen and found Gene planted in the aisle that led to the door. Two grown-ups could normally pass each other in that space, but not when Gene stood there.

Gene never came in the house unless he'd had plenty to drink, and this day he was ripped. He gave me his jerky, half-salute, forefinger cocked. He took deep breaths through his nose, letting the air out with grunts and whistles. He stretched out his hand. I had no choice but to take it. After what seemed like a few minutes of crushing my fingers, he mumbled something that sounded like "garbage."

"Sure, I'll check right away." I pried my hand loose and circled around the kitchen to the back door. Nothing in any of the cans. "Gene, there doesn't seem to be any garbage here. Sorry about that."

He turned to face me. He rocked back and forth, powered by his musical respiration.

"I get it. You've already collected the garbage. You want to be paid, right?"

He pointed his finger at me, looked at the ceiling, and after a moment he said quite clearly, "In a manner of speaking, y'might say that."

I scrambled around the kitchen again to avoid another handshake and went to the bottom of the stairs. I yelled for Chuck. I no longer had any part in paying the bills for Maple Tree Farm. Best to let him handle it, especially if it involved an overdue balance. Chuck came down and I hung back to listen without being seen. Gert came in to see what was keeping Gene. "When you broadcastin' these days?" she demanded.

"We're not doing anything right now," said Chuck. "Starting in the fall, we're going to do specials." He said this without hesitation. Perhaps he believed it. I thought about the transmitter antenna, which had collapsed unnoticed on the roof.

"What's a matter?" growled Gene. "You fallin' apart here?"

I didn't stay for Chuck's answer.

A few months later, I heard Chuck tell visitors we didn't have enough staff left to do Lanesville TV. The Council had funded Media Bus at half the amount we requested on top of a major cut the previous year. Lydia, our strongest supporter on the Council staff, had been purged, replaced with new staff who showed little sympathy for our style of video and no respect for our place in video history.

Each of us had applied for grants as individuals and we had received federal funds for the group, but the total remained far below what we needed as a group to survive.

Davidson and I had developed a private business building control rooms for a few clients–he did the carpentry and I did the wiring and handled the money. It didn't prove a reliable source of income. All of us were scrambling for work where we could find it.

We were all feeling the stress of not knowing where the money would come from to keep going. Just paying my own bills was eating into the remainder of my small inheritance at an alarming rate. I needed to find a job and I didn't know what I wanted to do other than video, nor did I have any idea where to begin looking. That didn't stop me from taking on big projects with no prospect of profiting from them. I used Lanesville as the staging point for a tape about a protest against a nuclear power plant under construction in Seabrook, New Hampshire, and convinced video groups from around the Northeast to participate. We used TVTV convention tapes and the *Five Day Bicycle Race* as models, with people agreeing to work on the project because they wanted to be there. They all believed in the anti-nuke cause, but the question, as always, was: What do we do with the tape once we've made it? There was no money for satellite time, and no major TV audience close to Seabrook. The answer I came up with was a scattershot grouping of cable TV public access channels throughout the region. As the demonstration wound down, a small armada of single engine, private planes ferried copies of our tape to the access channels in Albany, Boston, Kingston, St. Johnsbury, Vermont, and half a dozen other locations on our Northeast Video Network. The tape played simultaneously on those cable channels on the evening of June 10, 1978, and was updated at a prearranged time by a live telephone conference call from our headquarters in a dilapidated mansion near the demonstration site. It didn't play on Lanesville TV, because the station no longer existed.

The Northeast Video Network project left me exhilarated, exhausted and flat broke. I drove straight from Seabrook to Wall Street and an interview for a job as a techie at a big bank. In my interview I didn't hesitate to note I was the sole author of *The Spaghetti City Video Manual*, no matter what the cover said.

∞∞∞∞∞∞∞

Nancy and Bart decided in January of '78 they would move from Lanesville, but not until the fall. Funding cuts made it impossible to continue offering any sort of residential media center after that. Skip and Jane had left for San Francisco. Davidson showed up with his son, Murphy, to collect the last of his things.

Blackie had died, and whenever I traveled for the bank, I brought our golden retriever, Ratboy, to Sam's. He loved having a dog to walk, and sharing Ratboy was one way of maintaining the bond between us. He felt he was too old to have a new dog of his own.

I tried to explain the reasons why Media Bus had to move to Woodstock. I put the move in political terms, talking about a society that doesn't value its artists enough to support them, and about a bureaucracy that feared free expression through video. The truth was far more complicated than that, and Sam and Miriam must certainly have known it. "I am sorry you are moving out from mine house," he said to me.

The same ski club that rented Maple Tree Farm before we came had decided to take it again. Sam wanted "the boys," as he called the skiers, to stay over at the farm so they could get a head start on setting up their rooms. But Nancy wouldn't hear of it. The skiers, who were no boys chronologically, hung out by the back door, insisted on addressing her as "sweetheart," sucked on beer cans and made rude noises to display their impatience with the pace of our departure. She considered the farm her home until the day the lease expired, and she and Sam got into a shouting match about it. Sam worked himself into such a rage we feared for his health. Finally he went home, defeated.

A day later, he walked up Neal Road as we were tossing the last Media Bus gear onto a borrowed pick-up. His eyes had grown red and puffy, and the wrinkles that normally animated his face sagged like a hound dog's. He spoke in short phrases at first, but gradually became more himself as he found tasks to do in the empty rooms. Outside for a farewell photo, Chuck forced a comic pose, stuffing his fedora on Sam's head. Sam ignored him. He shook hands with his former tenants as I snapped a few photos and then went inside before they drove off. I followed him into the house. I didn't have anything to say, so I asked him to pose for a picture in the dining room that had served as our studio. He stared straight into the camera, frowning.

The heavily loaded pick-up headed down Route 214 for the new Media Bus studio and office on Tinker Street in Woodstock. The Media Bus board of directors, all the Videofreex but Chuck, and him, too, as a matter of course, had decided no one should live in the new space. The idea was to keep our private lives separate from our work.

Epilog

Between March 1972, when Lanesville TV went on the air, and February 1977, when we stopped broadcasting for good, the handwritten pages of the Channel 3 log record 258 different broadcasts. That averages one per week, although it didn't work out precisely that way, what with the early burst of energy for programming the station, and the periods in later years when the station went dark for a few weeks now and then.

Those 258 shows included 436 taped or specially produced live sequences, of which only about 15–or three percent–were repeated. Videofreex produced three-quarters of those 436 sequences. The rest were made by our hundreds of guests, except for a movie or two. I have these statistics because they are part of an official document issued by the Federal Communications Commission. The document describes Lanesville TV in some detail, and recommends that the commission adopt rules making it possible for the public to apply for licenses to set up completely legal stations along the lines of our Channel 3. I know this because I wrote that document. It's called *A Micro-TV Service in the United States*, and the federal government paid me reasonably well to prepare it.

In the spring of 1972, a young law school student from California named Michael Couzens had joined the TVTV crew preparing for the Democratic Convention. A number of Freex met him in Miami, liked his easygoing manner and invited him to visit Maple Tree Farm, which he did, several times. Michael and I had a passing acquaintance, but as activity at Maple Tree Farm wound down, I lost touch with him. I hadn't heard from him for several years, when he called unexpectedly in 1979 and said he was working at the FCC. Not only that, he now headed a commission task force charged with developing plans for a brand new Low Power Television Broadcast Service. Michael thought it would be a good idea to have a report on how Lanesville TV worked, as a kind of reality check for the government lawyers, engineers and economists assigned to the task force. He had an idea stations as small as Channel 3 wouldn't be practical; he had bigger stations in mind. But he knew how things worked in Washington, and he wanted me to use Lanesville TV as the basis for a written proposal for a lower-than-low-power TV service–a micro-TV service–just in case the commissioners freaked out at the prospect of authorizing new low power TV stations or had to bow to pressure from the financial interests of a broadcast industry fearful of new competitors.

If I agreed, would they try to bust me for participating in Lanesville TV? After all, the FCC Field Operations Bureau had been party to the raid in which Joseph Paul Ferraro was hauled out of his mother's house in handcuffs.

Michael had already thought of that and had secured the consent of his superiors to offer me immunity from my past extralegal appropriation of an unused space in the radio frequency spectrum known in the U.S. as Channel 3. So I said sure. I liked the idea of telling the government how we had ignored its rules, and I needed the money.

My research uncovered how pirate broadcasting actually originated in this country back in the late 1940s, when people out West, who couldn't receive TV any other way, set up little booster systems to retransmit the signals from TV stations in distant cities. In the beginning, they didn't bother with licenses, and when the FCC ordered them to shut down, the governor of one Western state threatened to arrest any federal agent who tried to block the rebroadcasts. In the end, the FCC blinked first. It authorized a limited form of station retransmission. The victorious pirates got licenses. They didn't see the FCC's decision as furthering or impeding their political goals. They just wanted to watch TV.

I also found a guy in Canada named David Brough, who was setting up tiny Lanesville TV-like stations in remote northern communities. When the Mounties confiscated the transmitter in one of these settlements, the entire town occupied the regional government building until they got their station back. After that incident the Canadian government quickly authorized what it called a Mini-TV Service.

And I met Dr. Frank Cyr, known as the Father of the Yellow School Bus, who built the first authorized low power TV station in the United States. He got his license in 1966, two years before Sony introduced its first portapak in the U.S., and he set up his station just a couple of mountains away from Lanesville.

Cyr was a nationally known innovator in education, and in the mid-1960s, he began to wonder how to bring educational opportunities, and particularly educational television, to isolated communities. He was living on the western slopes of the Catskills near the village of Stamford, New York, about a 45-minute drive from Maple Tree Farm. He knew most of the schools in the region were located in valleys, where they couldn't pick up the signals from educational TV stations in cities like Syracuse, Binghamton and Schenectady. He figured he could reach those schools with a centrally located transmitter rebroadcasting the signal from a distant station, but he didn't want to retransmit only one station. He wanted to pick the best programs each station had to offer, videotape them and then rebroadcast them at times convenient for the schools. He also wanted to produce local programming.

It was an ambitious project, but Cyr knew he could get federal funds to support it. His only barrier was the FCC, which had regulations forbidding exactly what he wanted to do. He couldn't afford to build a regular station, and the FCC bureaucrats told him there was no way he could set up anything smaller.

A few years later, we solved the same problem Cyr faced by simply ignoring the FCC and its rules. But that wasn't an option for a prestigious educator like Cyr. And his biggest problem was that time was about to run out on the availability of the funding for the project. He was desperate and needed help to get the FCC to bend the rules. So the resourceful Dr. Cyr picked up the phone and called Robert Kennedy, the junior senator from New York. Kennedy owned a home in the ski resort village of Windham, one of the communities Cyr's station would serve. That's all it took. Logjams at the FCC miraculously disappeared and the commission acted in record time, granting Cyr a waiver to put his station on the air.

By the time we arrived, the Stamford station and all its booster transmitters (one of them on the other side of Hunter Mountain from us) had become a hidebound institution, uninterested in experimenting with anything but the most tame and uninspired local programming. Cyr, who was elderly but whose critical intelligence was undiminished by his years, shook his head sadly at how conservative the station had become. He told me in the mid-1970s, when he was no longer in charge, that he thought the people who ran the Stamford system were a sorry lot, afraid to try the very types of community programming he thought would fulfill the promise that had led him to build the station.

We were never able to work out any sort of collaboration and gave up trying after a while. It was clear they didn't want to work with the likes of Videofreex.

In September of 1980 the FCC adopted Michael's recommendations for the new Low Power TV Broadcast Service, the first new broadcast service authorized by the commission since FM radio had been approved fifteen years earlier. It wasn't part of the official record of the rulemaking procedure, but before the proposal came to a vote, Michael handed the Field Operations Bureau a copy of my report. The FCC staff urged him to be sure there wasn't anything embarrassing in it.

I had told Joseph Paul I was writing the report, and we had agreed I would identify him only by his first two initials. And sure enough, the people at Field Operations expressed particular interest in one J.P., as named in my report. Michael feigned ignorance, so the gumshoes ran a check through their computer, looking for any broadcasting miscreant identifiable by those initials. The search came up blank. Then they checked the complaints file to see whether they'd heard from anyone about Lanesville TV. Again, nothing. I was clean. My report was printed in the Federal Record along with the new Low Power TV rules.

News of the adoption of rules for low power TV stations made the front page of *The New York Times*. The micro-TV service report and Lanesville TV didn't rate a mention.

Today, you may apply to the FCC for a license to build a low power television station in some rural communities. Figuring out what to do with a small station if you get a license is another matter altogether. And another story.

Parry Teasdale was a founding member of Videofreex. He authored *The Spaghetti City Video Manual*, and later produced independent documentaries and childrens' tapes for TV and video distribution. In 1979 he served as a consultant to the FCC, where he investigated the feasibility of small independent TV stations. After a few years in broadcast and cable TV, and in corporate video, he became the editor of *Woodstock Times*, a weekly newspaper in upstate New York. He and his wife, Carol Vontobel, also an early member of Videofreex, live in the Catskill Mountains. *(Photograph by Dion Oqust)*